The Biolab Book

VE-SAL-IUS
GALEN
DE PORCI CORPORIS FABRICA

The Biolab Book

Second Edition

Lundy Pentz

The Johns Hopkins University Press
Baltimore and London

Printed in the United States of America
94 93 92 91 90 89 5 4 3 2 1

The Johns Hopkins University Press
701 West 40th Street
Baltimore, Maryland 21211
The Johns Hopkins Press Ltd., London

The paper used in this publication meets the minimum requirements of American National Standard for Information Sciences—Permanence of Paper for Printed Library Materials, ANSI Z39.48-1984.

ISBN 0-8018-3707-3

Contents

Preface

Biology, by its very nature, ought to be one of the most fascinating courses a student can take—but too often it isn't. Lab courses, a chance to get students involved in doing things rather than reading about them, ought to be refreshing and exciting—but too often they aren't. It is in the hope of contributing a little to this need that I have penned this volume.

The object of this book is to set forth a series of more or less representative laboratory exercises suitable for general courses in biology, with a concern for their presentation in a stimulating way. It is written in a rather informal style and illustrated with my own pen-and-ink drawings, simply because that is what I have done for my students in the Evening College at the Johns Hopkins University, and now at Mary Baldwin College; they have enjoyed them and, I think, learned from them. In a lab course it seems appropriate to take advantage of the lesser formality of the situation, and I have done so.

I have avoided using unnecessarily technical language and have tried to make the questions really answerable from the exercises (as opposed to those requiring the consultation of reference books). In drawings, I have avoided the merely diagrammatic as not providing enough guidance, especially in dissections.

Above all, I have tried to present biology to my readers as I see it—a source of wonder and delight that there should be a world with such things in it. The more we understand about living things, the stranger and more wonderful they become. To show this wonder to others has been my purpose; if I seem to laugh too much, it is because wonder moves me so. Perhaps my more solemn colleagues will forgive me, for it is the same wonder that moves us all.

There are many people I would thank, but first of all I owe a debt to Dr. Michael Edidin, whose clear, elegant, and witty lectures inspired me and whose encouragement and help got this book under way. Mr. Anders Richter and Mr. Jim Johnston of the Johns Hopkins University Press were supportive and more than humanly patient during the writing process. And my wife, Ellen, contributed constantly with her professional advice, good sense, and deep understanding. During the revisions for this second edition I also find myself indebted to my daughter, Julia, who contributed as only a baby can by sleeping through the night at an unnaturally early age! I have also profited by the advice and ideas of two treasured colleagues, Dr. Eric Jones and Ms. Bonnie Hohn, who have been a resource and an inspiration for me all along. I hasten to add that any mistakes in the book are entirely my own contribution!

As the old scribes said, "Ego scripsi miser totius. Deo gratias!"

The Biolab Book

Introduction to Dissection

Dissection (please, dis′-section, not die′-section) is an ancient and respectable process engaged in by seventeenth-century gentlemen and all small children. It is simply taking something apart in order to see how it works. But be warned: this must be a careful process. If you don't know what something is, don't take it out until you have identified it! Many a once-beautiful organ, when hacked out of place, becomes so much unidentifiable hamburger. Although you won't

really be expected to put it all back together again, dissect as if you might have to. Look, probe, make drawings. Don't fall back on the old whine "I don't have any artistic ability!" Here is a secret: the real reason biology students are asked to make so many drawings is *not* that biology teachers are yearning to cover their walls with student art; it is that you never really see a thing properly until you try to draw it. Drawing forces you to see proportion, shape, texture, and a whole world of detail. Never let it slide by, and never try to fill in the details from memory. Above all, don't draw from the drawings in this book; draw from what is in your dissecting pan. Your pig may be different, and in any case, what you really see is what matters. The history of science is simply full of people dutifully seeing what they thought they were supposed to see instead of what was really there. Of course, if you say that your pig has two heads you must be prepared for a certain amount of skepticism; but there are such things as two-headed pig fetuses all the same. The world is full of things a great deal more peculiar than most textbooks care to admit, but you must teach yourself to see them clearly. If you don't, the world will be as rigid and orderly as a textbook—and as dull.

If you are going to describe the location of various parts of your pig, you will have to learn some technical terms. Just as on shipboard you must know "fore" and "aft," and so on, in the dissecting room you should know some directions also. For one thing, they are less ambiguous. Which way is toward the "front" of a pig? Your stomach is "in front of" your spine—but is a pig's? So take a deep breath and stand by for new words:

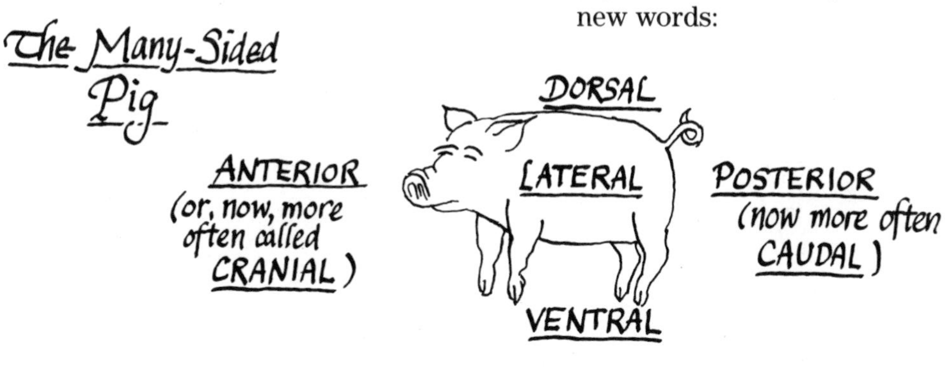

Now, you ask—what about left and right? Say just that, but always the *pig's* left and right, not yours. So you would say that the lungs lie in the anterior portion of the body and are lateral to the heart (or, to use the adverbial form, the lungs lie laterally to the heart).

You will find that you also need to refer to planes, and here also the usual way of speaking is ambiguous: is the diaphragm horizontal? When? Do you imagine the pig on its back in the dissecting pan, or as in life—or perhaps dancing? As you can for any three-dimensional object, you can slice a pig three ways:

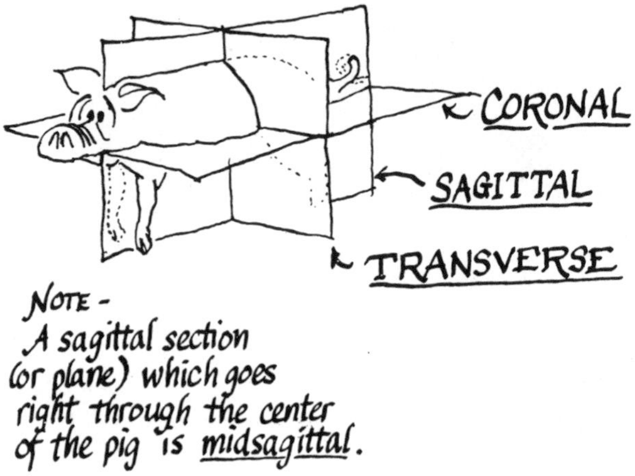

CORONAL
SAGITTAL
TRANSVERSE
NOTE -
A sagittal section (or plane) which goes right through the center of the pig is midsagittal.

One more small point, and you will know everything. For any two points along a limb, or a blood vessel or such things, the one closer to the body is *proximal* to the other; the one farther from the body is *distal* to the other.

Unfortunately, the anatomical terms for humans have a few differences from these terms. Here is a list to help you translate:

Human	*Pig*	*Human*	*Pig*
Posterior	Dorsal	Inferior	Posterior
Anterior	Ventral	Horizontal	Transverse
Superior	Anterior	Frontal	Coronal

Some writers prefer to use part or all of the "pig" terminology for humans also; you just have to know which they are using. Some day all will be standard. In this book, I will use only the "pig" terms, which are really used for all nonhuman anatomy.

How to Dissect

Dissection has its pitfalls, but its main pitfall isn't quite what you might expect. It is this: most people use the wrong end of the scalpel. Almost all scalpels are made with handles designed to pry apart clinging membranes or to free blood vessels from loose connective tissue. This technique, blunt dissection, is useful in many situations. It is most valuable for its gentleness. After all, you want to see things as intact units, not shreds. Dissection should be more like watchmaking than butchering! Pull apart before you cut apart.

And above all, *never* cut blindly into anything—be sure you know what's inside, or else cut in repeated, shallow strokes until you can see what's inside before you slice into it. That way you can have a good idea of how things really look, instead of having to attempt a reconstruction after the disaster. Often you will need to remove skin or enveloping membranes from something. Here is the best way to do it:

1. Make an opening by pinching up a fold of skin with the forceps and snipping a bit of its base with the scissors. This way you won't cut any underlying organs. Note that the forceps are not used to avoid touching things, but to give a firm grip on things too small or slippery for fingers.

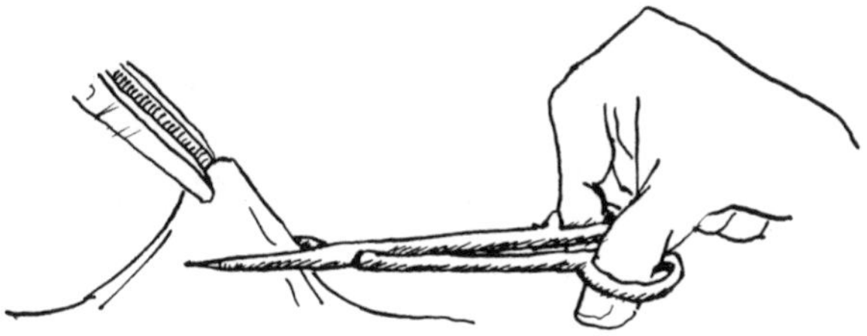

2. Insert the blade of the scissors into the opening and begin to cut, holding the scissors so that the skin is lifted away from the organs beneath.

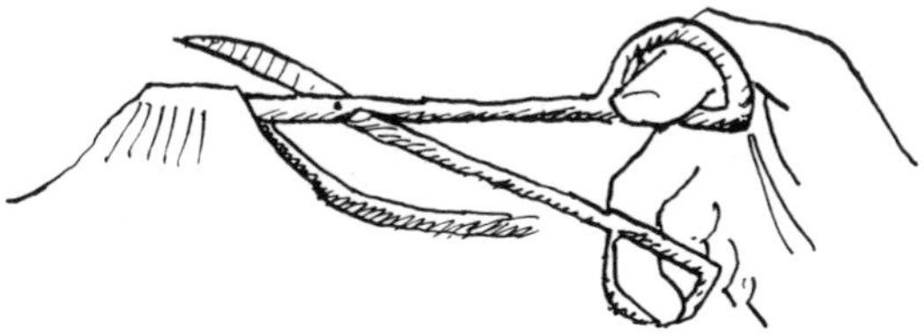

3. Using forceps or fingers, pull the skin back from the cut. Now, using a scalpel held flat, gently scrape free any loose connective tissue that seems to hold the skin in place. In some cases, you may need a fairly rapid alternation between (2) and (3) to get anywhere.

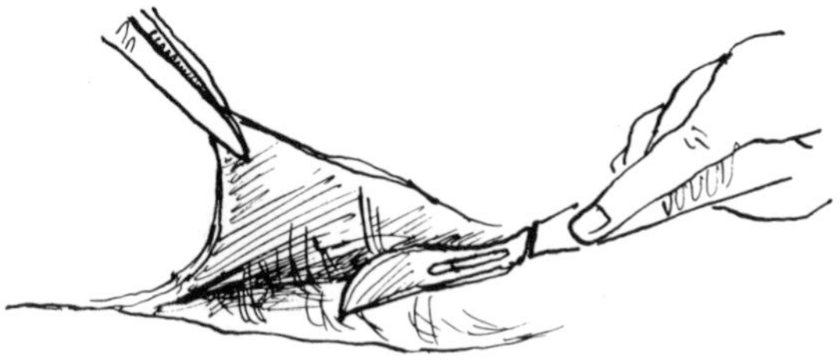

Now—to the Pig!

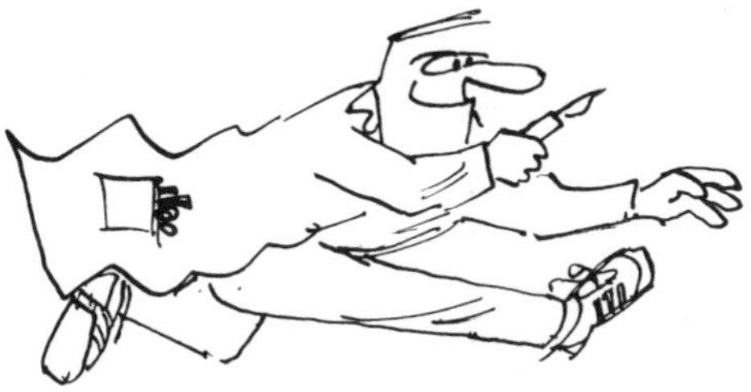

First of all, you may wonder: why a pig? Well, to be perfectly honest, one reason is that they are readily available and cheap! (You would be surprised at how much the choice of material for research purposes is governed by that.) But there are better reasons than this one. The fetal pig is a very convenient size to work with, is easy to preserve, and is a fairly typical mammal. In fact, its anatomy is quite like your own. This may not be the most flattering comparison you have heard, but it's quite true nevertheless. The real object of dissecting the fetal pig is to learn about mammalian anatomy in general, and this naturally includes human anatomy. It is a general practice to remark that the fetal pig lies in the dissecting pan in place of you yourself. Leaving aside the point that self-dissection is a difficult and dangerous practice, you will soon see the truth in that remark. Besides all this human interest, the pig can give you a view of several special fetal structures that made life in the womb possible.

Surface Anatomy

Get your pig and wash if off with tap water to remove excess preservative solution. Put it down in your dissecting pan and take a good look at it. What color is it? You may have noticed a soft brownish material peeling off when you wash it; this is the *epitrichium*, and it is an outermost skin layer covering hair and all other external features while the fetus is still in the uterus—sort of a disposable wrapper. Are the eyes open? Yet the newborn will find a teat and nurse greedily. Note the huge, rough tongue. Your pig will have a deep cut in the neck produced in the course of preparing it for dissection. How many nipples does your pig have? Does everyone's pig have the same number? Do you think you could guess the average litter size for a pig?

Now you want to determine the sex of your pig. This is not as easy as it sounds unless you know pigs. Don't waste your time looking for a penis; you can't see it. Look under the tail for a fingerlike projection, the urogenital papilla. If your pig has one, it is a female. If it does not have one, look between the hind legs (from the posterior) for a patch of skin much thinner than the rest; it may or may not contain testes yet, but it is the scrotum and indicates that your pig is a male.

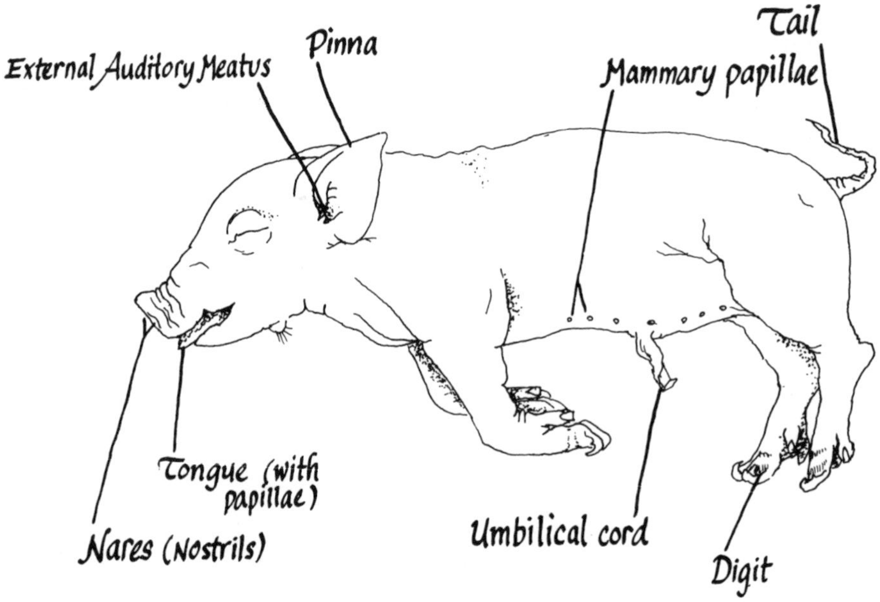

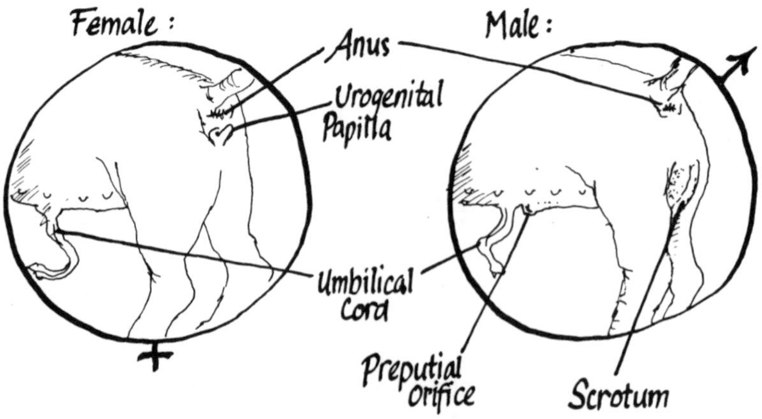

Opening Up

Before you proceed, be sure you know the sex of your pig; the incision is made differently for the two sexes. If you have sharp and fairly heavy scissors, use them in the procedure given in "How to Dissect," above. If you don't then you must use a scalpel; but do so very carefully, pulling apart the edges of the cut as you go. Begin in the abdomen and make the cuts shown. Note that for the male, you make two parallel cuts up to the umbilical cord; these should be 1–2 cm apart. The penis lies between them.

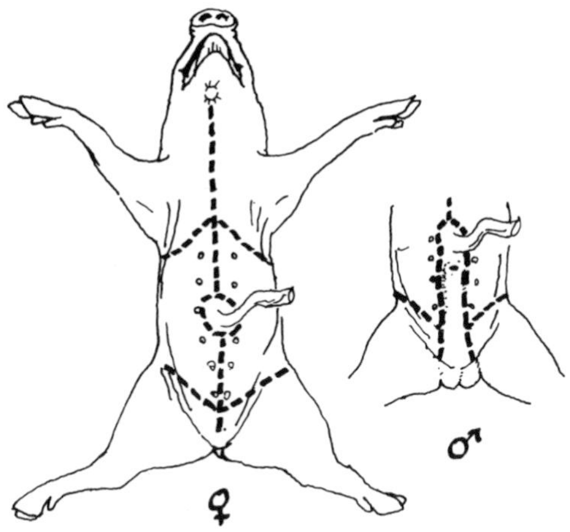

The cuts in the abdominal region should go through to a cavity, the *peritoneum.* But when you reach the thorax (chest), you will hit the rib cage. Peel the skin and muscle away from the ribs, but do not cut into the ribs yet. Make lateral cuts just posterior to the rib cage.

You may wish to use string tied to the pig's feet and passed under the dissecting pan to be tied to the opposite foot; this will help spread the legs apart. This is not, however, necessary. When you peel back the flaps, you should see something like the accompanying diagram. What probably strikes you first is the size of the liver. Of course, part of this is due to the pig's fetal condition, but the liver is a very large organ. Before going further, you should go to a sink and wash out your pig's abdominal cavity with lots of cold water. The brown material you wash out is mostly blood forced out when the blood vessels were injected with latex.

You will notice that the arteries are injected with red latex and the veins with blue. This is helpful for identification purposes but can be misleading. All that is blue is not a vein! You may also be troubled by big, shapless masses of blue latex resulting from a vein blowing up during the high-pressure injection of latex. Just pick these out and discard them.

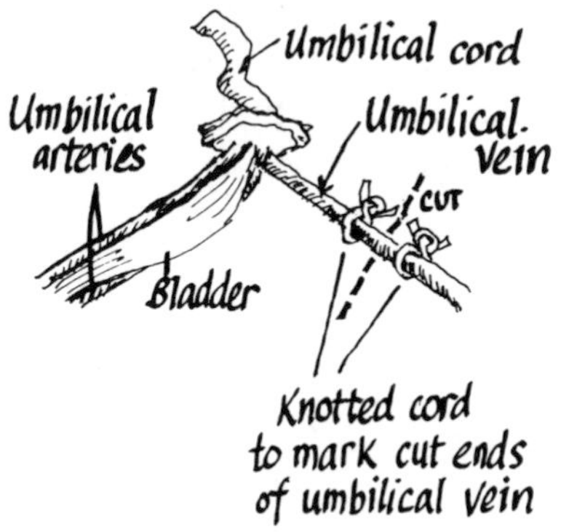

You will see at once why you left the umbilical cord on its own little island. It has anterior and posterior attachments. To get the cord out of the way, first tie two bits of string around the umbilical vein, and then cut it between them. Why the string? So that you can identify the ends of this important vein when you see them again. Now pull the umbilical cord down between the pig's hind legs.

Make a large drawing of the pig at this stage. Remember to include external structures like legs, so that you will have landmarks. Also, draw a bit of a centimeter rule, to give the scale.

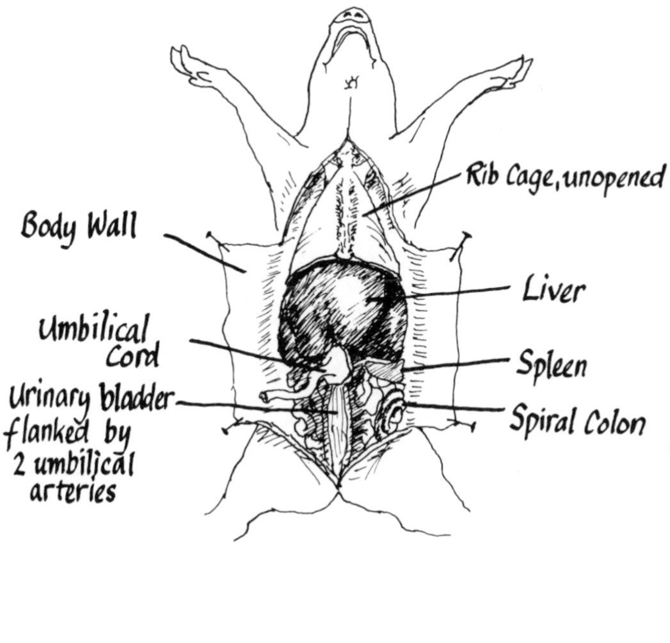

Cleaning Up

First, put your pig away in a plastic bag, with a little extra preservative. Next, wash your dissecting instruments with soap and water, rinse them well, and dry them thoroughly. Never, never, never put instruments away dirty or wet. They will become rusted and useless. Finally, wash your hands well. The preservatives aren't very dangerous, but you still shouldn't leave them on your hands. You may find them drying to the skin; applying cold cream before and after dissection will help.

Questions:

1. What special fetal structures have you seen so far?
2. Describe several structures in the pig in words, using the correct anatomical terms you learned in this chapter.

Organs and Systems in the Pig: One

Having opened up your fetal pig, you are now ready to study some of its parts. The way in which you study them should be governed by the architectural principle that form follows function—or, in other words, that when Little Red Riding Hood said, "My, what big teeth you have, grandmother," she shouldn't have needed the wolf's explanation of their purpose. Surely they weren't for decoration! When biologists observe a new structure, they immediately want to know what it's for, and the way it's put together should, they think, provide a clue. And when they have finally figured out what it's for, and how it does it, they expect to be able to explain its structural details in terms of its function. This doesn't always work out, of course, but it is generally so useful an approach that it is built into the very structure of modern biology. To show you this, I will constantly be talking about what things *do* as well as how they look—that is, physiology as well as anatomy. This of course, doesn't mean that biologists have had some mystical revelation to the effect that your every wart and freckle has a purpose. It just means that in practice, functional explanations make sense of things best.

Food—Getting It and Using It

Life can be divided into producers—mostly green plants—and consumers. Ever since the first consumer took a bite out of his self-sufficient neighbor, gathering and using food has been a primary concern for us. The pig is rather like humans in its truly broad-minded approach to food. It has, therefore, a digestive system like ours, a sort of intermediate type in between the short, simple tract of meat-eaters and the immense and complicated one of plant-eaters. We take whatever comes our way.

It all starts, once the food is caught, with the nose. Is this stuff safe to eat? Even humans, with our notoriously weak sense of smell, are regularly saved from ingesting bad eggs, dubious hamburger, and putrid fish by our tiny noses. Spend a moment contemplating the majestic shout on your pig and think what a virtuoso of smells the owner of *that* must be.

Now cut the corners of the mouth back to—and through—the jawbone, so that you can fold down the lower jaw as shown below. Some teeth will have appeared in older fetuses (one difference from humans), and you will be struck by the ridged appearance of the hard palate. This acts as a foil for the action of the tongue in rolling food into a ball.

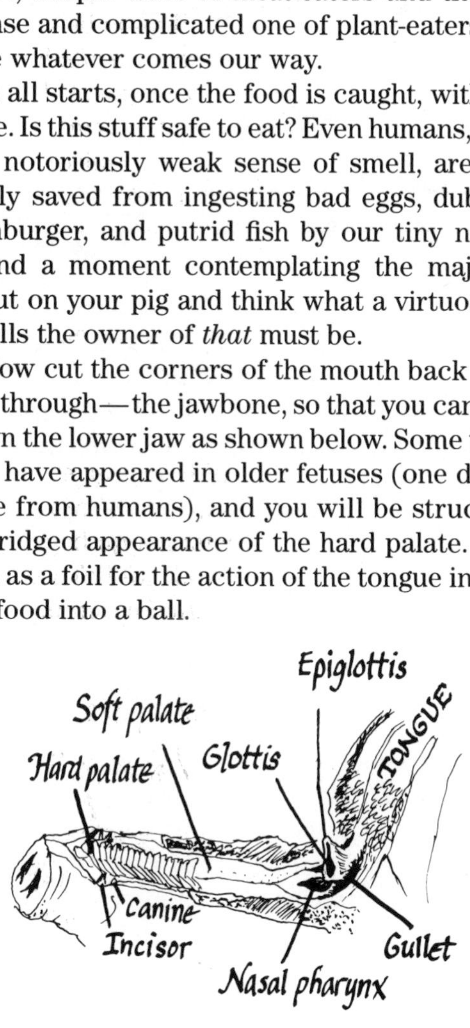

Toward the back of the mouth, you will see a projection rising from behind the tongue. This is the *epiglottis*, part of a clever system designed to keep the respiratory and digestive systems separate. The opening in its center is the *glottis*, and leads to the respiratory system; this opening, acting together with the opening of the *nasal pharynx,* forms a valve that closes when food is swallowed, to prevent it from entering the lungs. Food passes instead through the opening of the *gullet,* which surrounds the base of the epiglottis. So why, you ask, have the two systems connected at all? Well, have you ever had a bad cold? It is to any creature's advantage to have alternative breathing routes.

The food, having been chewed and moistened and rolled into little balls (*boluses*), is swallowed and passed along the esophagus (which we will not dissect until later) by a unique transport system. Each bolus is forced along your tract by circular muscles contracting behind it.

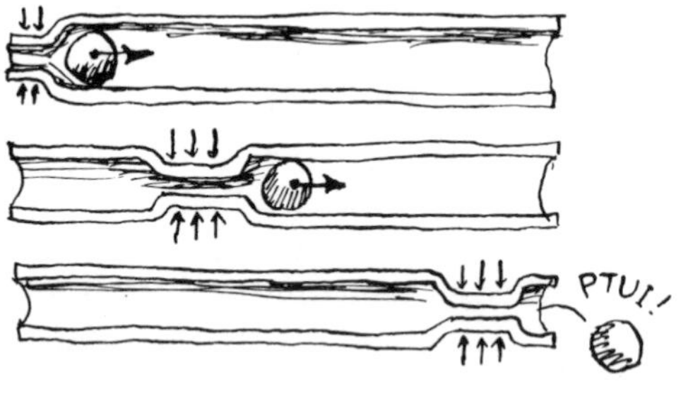

These contractions are regular and move in an orderly fashion like waves along the digestive tract. At least, they do when you're healthy. But viruses, food poisoning, and even tension can interfere with the orderly march of these waves, making them become erratic or even move in reverse—resulting in cramps, diarrhea, and vomiting. In the normal course of events, however, these peristaltic waves proceed at a steady rate—several per minute in the esophagus, becoming progressively slower until, in the rectum, they come only a few times a day.

To get a good view of the rest of the digestive system, you must first deal with the *liver*. This is unfortunately rather out of order, since as you know, the food goes next to the stomach; but such is life. Leave the food momentarily stuck in your throat and think about the liver. Such a versatile organ! For a long while, it is where the fetus's blood cells are produced (just at birth its blood-forming areas are shutting down). Throughout life, it is the site where many important blood proteins are made; it stores sugar and doles it out as needed, and performs a lot of other functions besides. But we can consider it as a part of the digestive system also, for two reasons. First, it produces bile—a bright green material made partly of recycled hemoglobin—and supplies it to the gall bladder, from which it is discharged into the digestive tract. There it acts like a detergent to break up clumps of fatty material for better absorption. Second, and even more important, the liver sops up all the nutrient-laden blood coming from your gut after a meal. It filters the blood through masses of cells which remove poisons from it, chemically altering them to harmless substances. This *detoxification* is essential, because many foods have potential poisons in them; but the process is indiscriminate and destroys medicines as well. Often the cells involved die in the process, and if enough of them die, a part or all of the liver may degenerate in various ways. Think of all those cells killing themselves to protect you from that martini or those tacos!

Anyhow, the liver is built rather like a sponge, as you might expect for a filter. If you shave away layers of it (making *coronal* slices), you will see one of more of the great, treelike blood vessels entering or leaving this mass of tissue.

Make sketches showing the vessels you see; note especially the umbilical vein. When you have done this, note the saclike *gall bladder* on the posterior side of the liver. Then remove as much of the liver as you must in order to be able to lift the stomach and fold back the intestines to the (pig's) right, as in this drawing. (You may be struck by the apparent lack of order in the intestines; those nice pictures of human anatomy with the large intestine making a neat square frame for the small intestine are fiction, as any radiologist will tell you.)

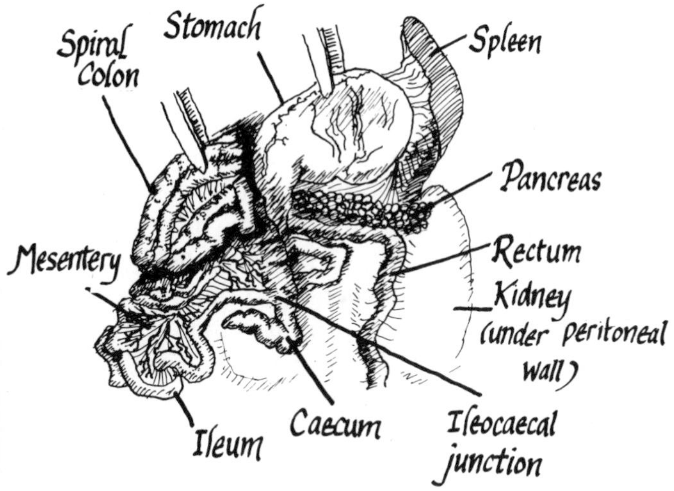

The stomach is a big, slack bag; feel its posterior end for a thickened area. This is the *pylorus*, a *sphincter* that keeps food in the stomach from going too far, too fast. Sphincters are rings of muscle which can contract to close a tube. (Another way of doing the same job is with a *valve*, a flaplike affair found mostly in the circulatory system.) Remove the stomach and cut it open from the esophagus to the pylorus; wash out the greenish *meconium* (mostly sloughed cells and bile) and look at the glistening white lining. It is thrown up in long folds to allow for expansion like pleats, and is coated with mucus to protect its delicate cells from being damaged by the hydrochloric acid they secrete. Why acid, you say, especially when it percolates up your esophagus and gives you heartburn? Well, for one thing, your stomach produces digestive enzymes, which need acid to work well; but that's really begging the question, since there are also lots of enzymes that don't need acid. Perhaps the real reason has to do with the fact that most bacteria are not able to get along in an acid environment.

The next bit of plumbing downstream from the stomach is the *duodenum* (dew-oh-dean′-um); it is here that all sorts of things begin to happen. As the food enters it, the duodenum responds by dumping a hormonal signal into the *pancreas*, the rather loose, knobby-looking organ dorsal to the stomach. The signal provokes the pancreas to release a solution of bicarbonate (baking soda) to neutralize the stomach acid; this solution also contains a large number of enzymes which are responsible for most of the digestion of food. (The enzymes in the stomach only begin the process.) This complex solution is released into the duodenum through a short tube, the pancreatic duct; and if a meal contains fatty substances, bile will be released also, through the longer and more visible *common bile duct*. This is a junction of two ducts, one coming from the gall bladders and one directly from the liver.

The rest of the small intestine (the *jejunum* and the *ileum*–the anterior 2/5 and the posterior 3/5, respectively) puts the last touches on the breakdown of food and absorbs the nutrients from it. You may have noticed that the rate at which an audience can leave a theater has a lot to do with the number and size of the doors. In just the same way, the efficiency of nutrient absorption depends on the surface area of the tissue doing the absorbing. So the great length of small intestine shouldn't surprise you. Your own is three meters long—twice that when totally relaxed, as in a cadaver. Look at the thin *mesentery* supplying the intestine with blood. You will have to stretch out a loop of intestine to see this. Here, fine arteries bring blood into the intestinal wall, and fine veins (probably not injected in your pig) carry nutrient-laden blood out to be passed through the liver. Cut out a short section of small intestine and cut it open along one side; wash it out and spread it, inner side up, in a dish full of water. Shine a good light on it from a very low angle and inspect it under a dissecting microscope or with a hand lens. See! We told you about surface area. All those fingerlike projections, *villi* (vil′-eye), are ways to increase surface area without increasing the length of the intestine. The whole surface area of your intestine is equal to that of a football field! It's so efficient that almost all the nutrients are absorbed in the jejunum, leaving the ileum to pick up stray vitamins and reclaim the bile. Each little villus grows from the base and sloughs off the tip (giving you a sort of perpetual internal dandruff, which accounts for a fair bit of your stools), and each one has a little slip of muscle inside so that it can wave about gently in the sea of rich broth which bathes it.

We now come to the less glamorous end of the tract, the *colon*, or large intestine. In the pig this is, for the most part, a tightly wrapped spiral mass, the *spiral colon.* It begins at the *ileocecal junction,* where there is a sphincter to prevent backflow. Cut open the colon and look for this sphincter; it protrudes like a finger into the colon. The short, dead-ended part of the colon on one side of this junction is the *cecum* (seek′-em); animals that eat a lot of cellulose—for instance, grazing animals—keep a colony of bacteria here, which ferments the cellulose. In humans, the cecum is short and bears the infamous appendix on one end. In the other direction from the junction, the colon goes through its spiral contortions, loops through the duodenum, and then makes a long, straight run (the *rectum,* meaning "straight thing") down the dorsal body wall to the *anus.* The colon is host to an immense colony of bacteria, some of which can cause serious diseases anywhere else in the body. They live on the indigestible residues of food and sloughed cells, and generally keep up a steady fermentation, sometimes generating very distinctive aromas. A change in diet or water can interfere with them, and antibiotics—particularly oral ones—can decimate them. Only when they're gone do we really appreciate how important they were to our general health. It used to be said that they produced certain vitamins that we require, but this is of dubious importance; it seems that the favorite animal for such studies, the rat, has the habit of eating some of its own feces, thereby getting the bacterial vitamins into the stomach and small intestine, where they can be absorbed. Unless this practice catches on with humans, we really can't claim our colon bacteria as a source of vitamins. They do form a delicate, balanced community, however, whose continued presence is somehow very important.

The main business of the colon is the reclamation of water. From the first squirt of saliva on, you moisten your food more and more, until it becomes quite soupy. If you just passed this soup along you would have to drink an extra 10 liters of water a day to replace your losses. Instead, you reabsorb the water; much of this is done in the ileum, but the last part is done in the colon, which removes all but what is needed for efficient export. Usually you don't overdo this,

but once in a while overdehydration takes place and constipation results. The consequences are not nearly as grave as the laxative-pushers would have us believe, unless they are made so by anxiety and laxatives!

Go back now and survey this system in your pig—this linear processing plant for food. Its muscular coats propel the food, its muscular sphincters retain it, and a system of changing acidity and various enzymes breaks it down step by step. It is all coordinated by signals, both nerve impulses and hormones. And by all this, the complex chemicals that spell *corn* or *squash* are disassembled into letters that can be used to spell new words like *pig* or even *you.*

Questions:

1. Complete your drawings and be sure to label all parts underlined in the text.
2. Sketch a straightened-out digestive tract, lable the sphincters, and summarize what happens in each division created by the sphincters.

Note: Remember to see that your pig is moistened with preservative and sealed in plastic before you put it away.

Organs and Systems in the Pig: Two

Now that you have been through the digestive tract (so to speak), you should be ready for anything. You have followed an entire system from beginning to end. Now you will get a look at several systems that you can't follow so completely (for example, because they become microscopic at some point).

Circulation and Respiration

One of the consequences of being big and complicated is that supply and distribution become problems. It is always a shock to new outing leaders or convention hosts that dinner for 100 is *not* simply 100 × dinner for one. If all your billions of hungry cells had to line up to get their dinner, the effect on your internal organization would be devastating. It is much better to give them room service.

One way to provide the room service is to introduce a nightmarishly complicated plumbing system that delivers the goods (and removes the garbage) in liquid form. Everything is brought to your cells so closely that no cell is more than a few times its own width from the supply line. The proper name of this sytem is, of course, the *circulatory* system, and you will begin with its pumping station.

First, open the *thoracic cavity* by cutting the ribs laterally. Use good scissors and if they have a blunt point direct it inward. It is easiest to begin at the posterior end. Pull away from the underlying organs as you cut. You will also need to cut the *clavicle*, or collarbone, which is ventral to the ribs, in much the same location as your own. Finally, cut through the *diaphragm*, the muscular wall separating the abdominal and thoracic cavities; cut it next to the ribs and lift off the rib cage; then peel back the sides of the neck incision, and your pig should look like the one at the top of the next page. The heart is swathed in an almost transparent membrane, the *pericardium*. It will be partly hidden by a huge, dull-colored mass of tissue whose surface is broken up by fine fissures. If you look up into the neck, you will see two fat fingers of the same tissue there, and (unless you broke it removing the clavicle) a thin bridge connecting the three in the middle. This isn't fat, as you might think. It is the *thymus*, an organ that produces mature, specialized white blood cells of a certain kind. These cells will patrol the body and remove infected or defective cells; they are probably important in protecting you from cancer, as well. The thymus is the place where these cells become specialized to do this; it is full of cells in the fetus, and in the growing child it shrinks. Yours isn't a great deal larger than this pig's, and by the time you're 30 it will be smaller and mostly fat.

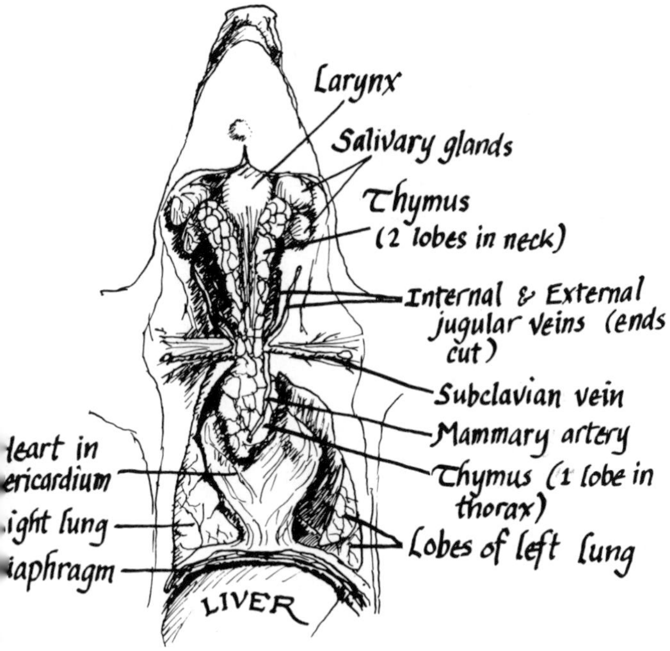

Remove the thymus and carefully peel away the pericardium. The resulting view should resemble the one below: the most prominent feature (besides the heart itself) is the great, branched vein that enters the anterior end of the heart. Veins are named in a way that seems far from obvious; unlike rivers, their main stream changes its name every time a tributary joins it. The main trunk of this vein is the *anterior vena cava*, but its branches, which divide immediately, all have different names, as in the drawing.

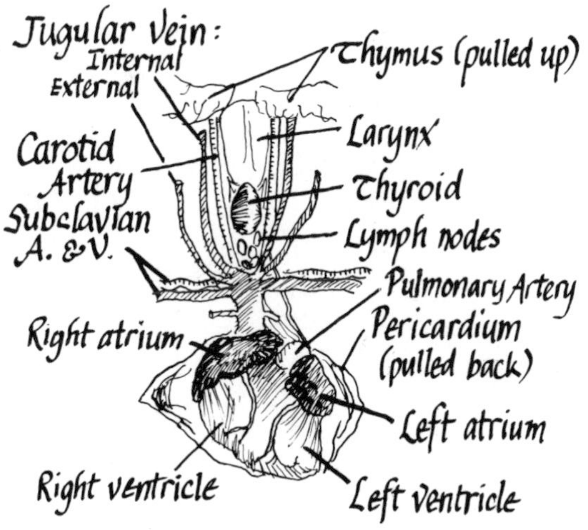

If you peel away a thin coat of muscle, you can see the pearly cartilage of the *larynx* (lair′-inks), or voice box—compare its shape to that of your own, the "Adam's Apple" you can feel in your throat. Below it lies the *thyroid* gland, the source of hormonal signals that regulate the intensity of your body's chemical activity. Just above the point where the vena cava branches, you may see a cluster of little whitish bumps; these are *lymph nodes*, clusters of white blood cells which become enlarged and active when you have an infection. They are found all over the body but are most prominent in the armpits and groin, where they are mistakenly called "glands."

Look at the *heart*—you will note that while it is not as far to the left side as melodramatic actors seem to think it is certainly tilted that way. It is, as you expect, tough and muscular but what are those delicate, dark-colored caps on it? They are, in fact, two of its chambers the right and left *atria* (ay′-tree-uh, pl; the sing.: *atrium*)—the name is that of the foyer of an ancient Roman house, and these chambers are the ones through which blood *returns* to the heart. They are a relatively low-pressure system, so no thick walls are needed. They have appendages, auricles, "little ears"; this gives a rather good description of their shape and position. Posterior to them are the *ventricles*, the high-pressure system that pumps blood out of the heart. Their wall is thick, made up of many layers of muscle fibers criss-crossing each other like the reinforcements in a radial tire. Here the force is applied to form the surges of blood you perceive as your pulse. The origin of the signal for these rhythmic contractions is in the heart itself, more particularly in a little group of cells located just where the anterior vena cava enters the right atrium. If this natural pacemaker is damaged or weak, the various muscles in the heart keep beating but at their own pace. To organize this, an artificial pacemaker is installed.

Remove the heart. To do this, cut through the anterior vena cava, pull the heart to the (pig's) right, and cut the pulmonary artery, aorta, and pulmonary vein on the left side. The pulmonary artery makes a loop over the heart and then goes in a posterior direction, where it branches to the right and left. You may have to cut it in two places to free the heart. Then pull the heart to the left, cut the right pulmonary artery and vein, and finally pull the heart anteriorly and cut the posterior vena cava. You should now be able to free the heart completely and lift it out. The cavity should look like the drawing below. (Depending on whether or not you cut above a branch point, you may not see some vessels.)

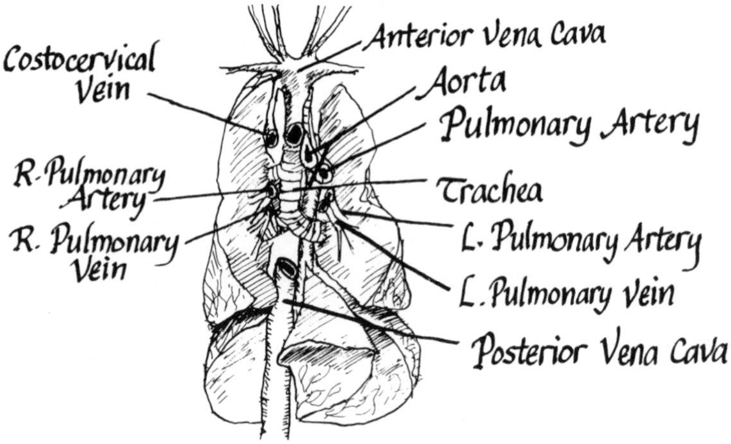

Take the heart you have removed and look at it closely. In addition to the major vessels entering and leaving it, its surface is traversed by small arteries and veins in pairs (though in your specimen the arteries may be the only ones injected with latex). These are the *coronary* arteries and veins, and their job is—strangely enough—to provide the heart itself with blood. Its walls are so thick that the blood inside is too far away to supply most of the muscle cells. So crucial is the blood supply to these muscles that any block in their flow results in damage to the heart; if the block injures the heart enough, a heart attack, or "coronary," results. The name arises from the fact that the vessels encircle the heart like a crown (*corona* in Latin).

Slice the heart open in a coronal section (the connection of this with crowns is not clear!) and pick the injected latex out of the chambers. Behold! A self-healing, high-powered pump guaranteed to last a lifetime! Blood enters the right atrium, largely because it is pushed there as muscle contractions all over the body squeeze blood out of your veins and up toward your heart. (Now you know why you'd rather walk a mile than stand for that length of time!) As the heart relaxes between beats, the blood then enters the right ventricle. But on the next contrac-

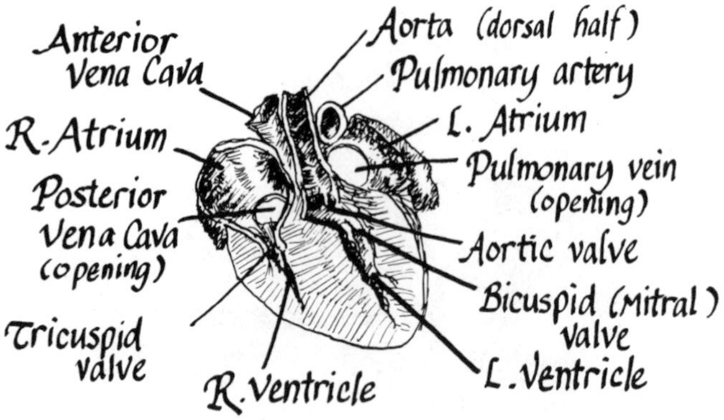

tion it is forced violently against the other side of the tricuspid valve through which it entered the ventricle, and this side won't give. It seals up firmly and is prevented from blowing out into the atrium by the *chordae tendinae*, fine anchor lines which you can see running almost the length of the ventricle (pry back the sides of the chamber to see them).

So the blood goes through the only available opening into the pulmonary artery. From here it enters the lungs for gas exchange, returns to the left side of the heart through the pulmonary vein, and is ultimately ejected through the aorta by a system exactly parallel—both literally and figuratively—to that described for the right side.

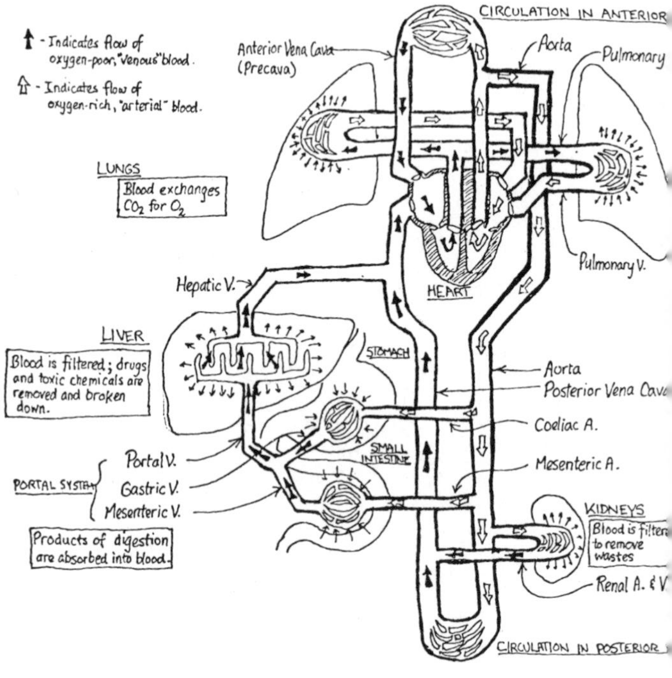

Where does it go from there? Round and round, as William Harvey found. Out through those thick-walled arteries, branching and getting finer and finer, at last the blood arrives in capillary beds, gossamer webs where it can get close enough to swap molecules with nearby cells, supplying their needs and removing their trash. Then a return is made, more or less efficiently depending on your state of activity. You should easily locate all of the vessels named on the diagram.

In the fetus, the source of nutrients and site of gas exchange is the *placenta.* The umbilical arteries and vein supply depleted, waste-laden blood to it and return oxygen- and nutrient-rich blood from it. Look to see where these vessels connect to the rest of the circulatory system. Since the lungs aren't needed, they are effectively bypassed; a passage between the atria—the *foramen ovale* (fo-ră′-min o-valley′), which can be located by probing gently just dorsal to the aorta within the atria—permits blood returning to the heart to pass directly into the left heart without going through the lungs. The *ductus arteriosus,* a short bridge connecting the pulmonary artery to the aorta near their point of emergence from the heart, allows blood pumped out of the right atrium to enter the aorta rather than continuing toward the lungs. When the lungs first inflate with air at birth, their greater resistance to blood flow sets off a rapid sequence of events that closes off these short circuits.

The *lungs* in your pig are unremarkable looking. Like ours, they are divided into lobes; if you slice into a lobe, you will see its central tree of arteries, veins, and *bronchi* (bronk′-eye)—the last are the smaller divisions of the *trachea,* which has carried the air from the larynx on down. Notice the hard rings on the trachea and bronchi—they are cartilage, and serve a function much like that of the ribbing in a vacuum cleaner hose: they keep the tube from collapsing under low internal pressure. (Here is one difference between pigs and humans—your pig has two branches of the trachea, or primary bronchi, entering its right lung; you have only one.) You probably know how the respiratory system works, the expansion of the closed thoraic cavity creating a low pressure inside the millions of sacs in which the bronchi's branches end, causing air to flow down the trachea to fill them. But, as the capillaries are the remote end of the circulatory system, so alveoli are of the respiratory system—details vital, but so tiny that we will need the microscope to see them.

Questions:

1. Using a bit of string, trace the path of blood through the heart. Ignore the fetal modifications. Draw the heart with the string in place.
2. Do you notice any consistent difference between arteries and veins in your pig—besides the color of latex, of course!—and can you relate this to function?

(*Note:* Tired of dissecting? Want a diversion? If you have compressed air lines—they look like gas jets but say "*Air*"—or a *big* rubber bulb, hook the air source to a rubber tube, the tube to a medicine- dropper tube, and insert the nozzle into a cut made in the trachea, pointing the nozzle anteriorly and holding it firmly. Then let in a burst of air—if you do it right, the vocal cords will vibrate and your pig will squeal loudly!)

Nervous System

The coordination of a nervous system allows the organism to respond to challenges as an integrated whole. It's a terribly important system, but its subtle electrical and chemical workings leave little for the dissector to see. On the left side of the pericardium, you will notice a thick white strand running down toward the diaphragm. This is a fairly typical nerve, the *phrenic* nerve. This particular nerve (one of a pair) travels from the spinal cord to the diaphragm, and its impulses cause the movement of the diaphragm in breathing.

Sensory organs are means of contact with events in the outside world. They are as varied as the events themselves. Most are microscopic, but one, the eye, is easy to dissect. Cut the skin at the corners of one eye, as if enlarging the slit formed by the closed eyelids. Then, at the far end of each cut, make a cut at right angles to it, so that the final pattern is an *H*. You can now peel back the eyelids and expose the eye. Its cornea will be cloudy—a change that sets in with death and is increased by preservation. Remove the eye from its socket by running your scalpel blade around the edge of the opening until you have cut the thin muscles that hold and move the eye; then reach underneath the eye and sever the optic nerve. Lift the eye out and, using a smooth, firm motion with a sharp blade, cut it in half along a line between the center of the cornea and the optic nerve. The *vitreous body*, like a thin jelly, fills most of the eye with a glassy clear material. The fat *lens*, opaque in your preserved pig, focuses light on the delicate *retina*, an array of cells which responds to light by forming nerve impulses. The retina is a thin (and now probably wrinkled) membrane, here easily seen against a black background—look at the darker area around the cornea, where the retina stops, then travel toward the back of the eye. (The black color you see appears at death—in life it is orange.) The star-shaped area on the retina is the spot where nerves and blood vessels exit from the eye; here no impulse is generated in response to light, so this is the *blind spot*.

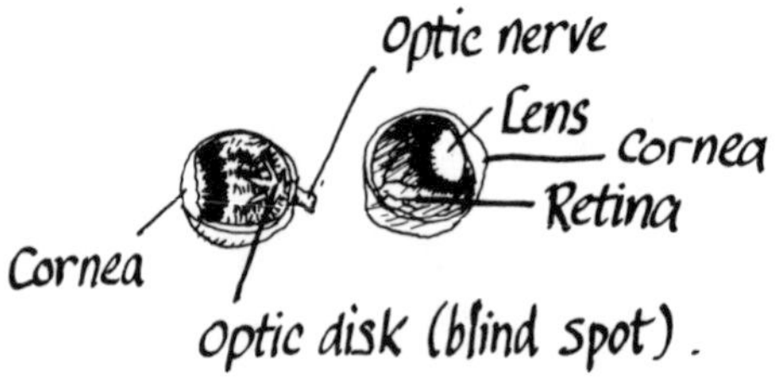

The master control is, of course, the brain. To see it, you must get a pair of sharp scissors. First, scalp your pig—remove a tapering square of skin between the eyes and the base of the ears. Then cut away the top of the skull in a similar way, being particularly careful not to poke inside. You will find the brain covered by an unusually tough, parchmentlike membrane, the *dura mater* ("hard mother"). (Well, you wouldn't want your brain in a flimsy bag, would you?) There will be a big vein, actually a blood sinus, running toward the back of the head under the dura, and then the brain. Peel away the dura and look at the convolutions of the brain. This is not the formless mass it seems—it is an incredibly intricate mesh of electrically active cells ar-

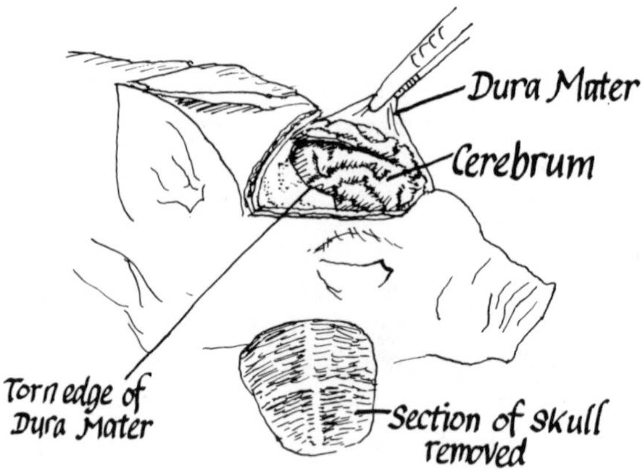

ranged in layers. To get more area for the layers without having a huge hollow brain, a wrinkled surface is needed. To some extent, different jobs can be found for different convolutions (through this varies), but bulges and bumps are signs of sickness, not character or intelligence. Einstein's brain looked just like yours or mine.

Question:

1. Draw what you can see of the brain's blood supply. Why is this important?

Reproduction

We can't leave our fellow mammal without a look at the reproductive system. Like us, the pig is set up for the intricate and rewarding maneuvers required by an entirely internal system of begetting and nurturing new life. Do the dissection appropriate to the sex of your pig, and look at one of the other sex.

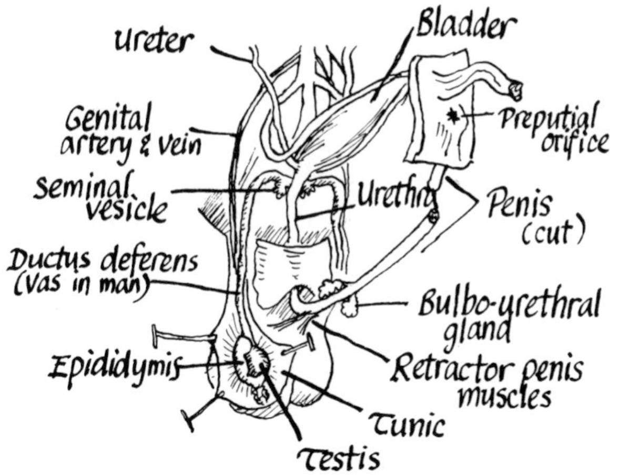

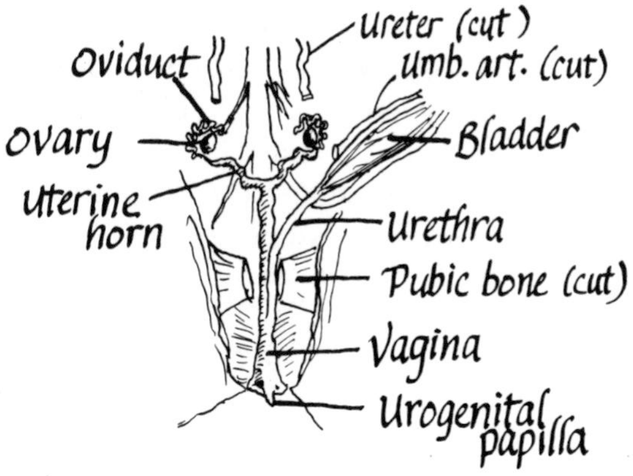

Male: Tie off and cut the rectum and move the intestines to one side. Find the *preputial orifice* on the skin posterior to the umbilical cord; feel under the skin to locate the *penis* (quite thin and long) and dissect it free from the skin. Cut through it and look for the three bodies of erectile tissue. Follow the shaft of the penis down to the *pubic bone;* cut this on each side of the penis and remove it, and find the *urethra* passing under it. Open the *scrotum* and you will find each *testis* enclosed in a tough, white *tunic*; the *epididymis* collects sperm and probably aids in their final growth; the *ductus* (or *vas*) *deferens* conducts sperm to the urethra. It is cut and tied in a vasectomy. You will see *seminal vesicles* in many animals, but in mammals they are *not* storage places for sperm. They, along with the bulbo-urethral glands and the prostate, which lies between them but is tiny and hard to see in the fetus, provide the fluid part of semen.

Female. Tie off and cut the rectum and move the intestines to one side. Spread the legs and cut in the midline between them with a sharp scalpel. You will reach the pubic bone; cut *just* through it and spread the legs firmly. Cut through one umbilical artery and pull the bladder to one side. (You may need to cut one ureter also.) The *ovaries*, smaller than the testes, are surrounded by the convoluted *oviducts* (fallopian tubes in woman), which gather up eggs (*ova*) and are the usual meeting place of egg and sperm. Once fertilized, the egg passes into the *uterus*, which in most animals that have litters is divided into two branches, or *horns*. In your fetal pig, the uterus is a pale shadow of the muscular thing it becomes in the adult (for that matter, these demure, smooth ovaries are hardly exemplary of the rough, pimply-looking mature specimens). The long, straight, thick tube distal to the uterus is the *vagina* (accent on the second syllable). The *urethra* merges with it in a vestibule; at the ventral edge of the vestibule is the *urogenital papilla* (perhaps to give directional guidance to a clumsy male?), and at its base, all but invisible, the *clitoris*.

Questions:

1. In some mammals, such as man, the ovaries discharge eggs directly into the abdominal cavity and the oviducts catch them, so to speak, on the fly. In other mammals, the ovaries are enclosed in a membrane and the eggs are more limited in motion. Can you determine which category your pig is in? (Check someone else's pig if yours is male.)
2. The testes develop in the same position as the ovaries, and then, near birth, they are pulled into the scrotum through a hole in the body wall, the inguinal canal (one on each side), by the shortening of a connection, the *gubernaculum,* which you can see just distal to the testis. Can you return the testes through the reverse way? Clip the gubernaculum and try. (This sometimes happens with very small, very cold children.)

The Microscope

The microscope is an instrument that has become as characteristic of biology as is the stethoscope of medicine; but it is also the holy terror of students. One wonders how many generations of students have obediently gazed at what Thurber called "lacteal opacity" and dutifully supposed they were seeing microbes or whatever. The ability to pretend you're seeing what you're supposed to see can get you through some classes, but it invariably results in boredom, headaches, and the conviction that microscopy is a rotten way to spend an afternoon. What a loss! The early microscopists nearly went mad over what they saw, and the sheer beauty of the microscopic world has stimulated generations of artists and photographers. (Look at some of the work of Roman Vishniac, for example.) If you use it properly—and this does take practice—the microscope can be as clear and as easy as a window, and the world you see through it can draw you into its spell like an enchanted wood in an old story. Beware!

Background: Lenses

Light, as we all know, travels in straight lines—that's what enables you to pick up your change from the store counter, start your car, and drive away without mishap—you can tell where things are from the light they reflect. And, of course, this is true—except when light travels through different substances. Then its path is bent; the most familiar example being the way the bottom of a body of water, or objects on it, seem deceptively close to the surface. The light which is reflected from the object bends as it leaves the water and enters the air, and you, assuming that it travels only in straight lines, misjudge the distance. People have exploited this fact for centuries, usually using a glass-air boundary rather than water-air, and a curved boundary rather than a straight one. The glass is somewhat more permanent than water, and making the surface curved causes the light to be gathered to a point—and to pass the point and widen out to infinity, carrying a steadily enlarging image. (For instance, the farther away you move the slide projector from the screen, the larger will be the image on the screen.) Such a curved glass-air surface is a *lens*. The grinding of lenses went from a trade to a hobby, and the country gentlemen took it up. It became fashionable at the royal court to carry a magnifying lens with which to examine your friends' fleas.

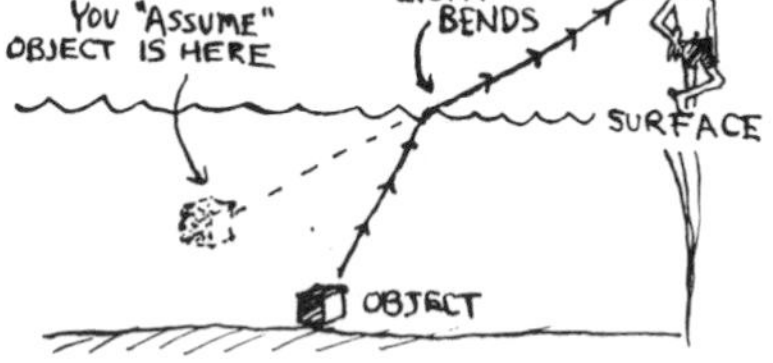

Background: The Microscope

People are never satisfied. Gentlemen in the seventeenth century wanted to magnify their fleas more and more. "Why," they asked, "can't you use one magnifying glass to magnify the image produced by another?"

Don't laugh; it worked! That is just what a compound microscope is: one lens magnifying the image produced by another! There are two main lenses, really systems of lenses: the *objective*, which produces an image magnified anywhere from 4 times or less to 90 or 100 times (a "real" image, which can be seen clearly if thrown on a screen), and the *ocular*, which enlarges this image anywhere from 5 to 15 times. The ocular produces a "virtual," not a real, image, one that cannot be seen clearly on a screen; it lacks a final lens element, one that is, in fact, the lens of your eye, which converts the virtual image to a real one thrown on the screen of your retina. Some microscopes also have a *condenser* lens system to shine a very concentrated beam of light on the object being viewed; this is important at high magnifications.

The idea of the two diaphragms (which are openings of variable size, just like the stops of a camera) is to reduce glare, which is caused by stray light bouncing in at the edges of the system. The slide is, of course, clear glass and rests on a support, or "stage," with a hole in the middle to let the light through.

There are two basic microscope designs found in most student labs; by far the most common is the *straight* design. It may be as simple as the one shown below, or it may have a condenser, built-in lamp, and mechanical stage. The other design, an *inclined* microscope, may not have all the accessories shown. The big difference between the two is this: in focusing, the straight scope moves its body tube, while the inclined scope moves its stage. The same movement that raises the tube of a straight scope will raise the stage of an inclined one, with the opposite effect. Throughout this and other exercises, the language appropriate to straight scopes will be used; if you have an inclined one, you will have to interpret: "raise the objective," to you, will mean "*lower* the stage."

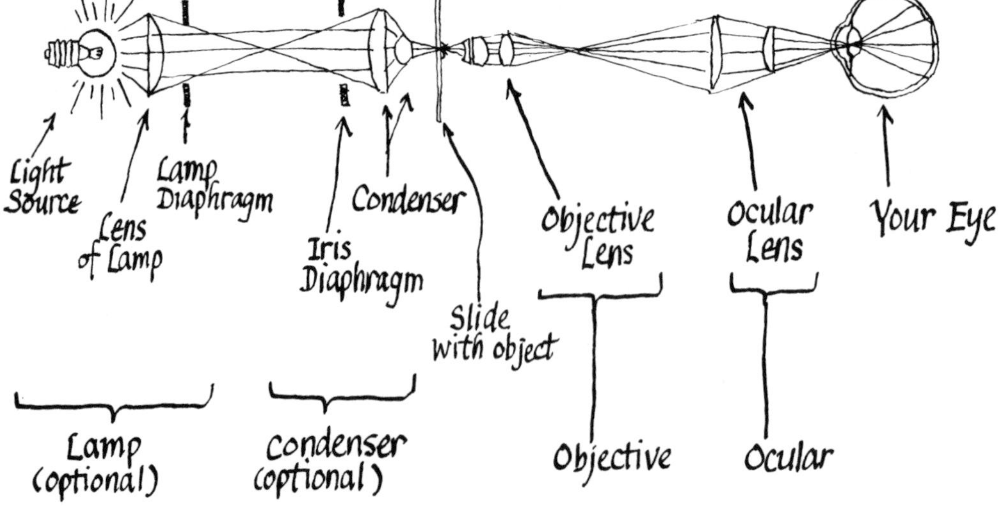

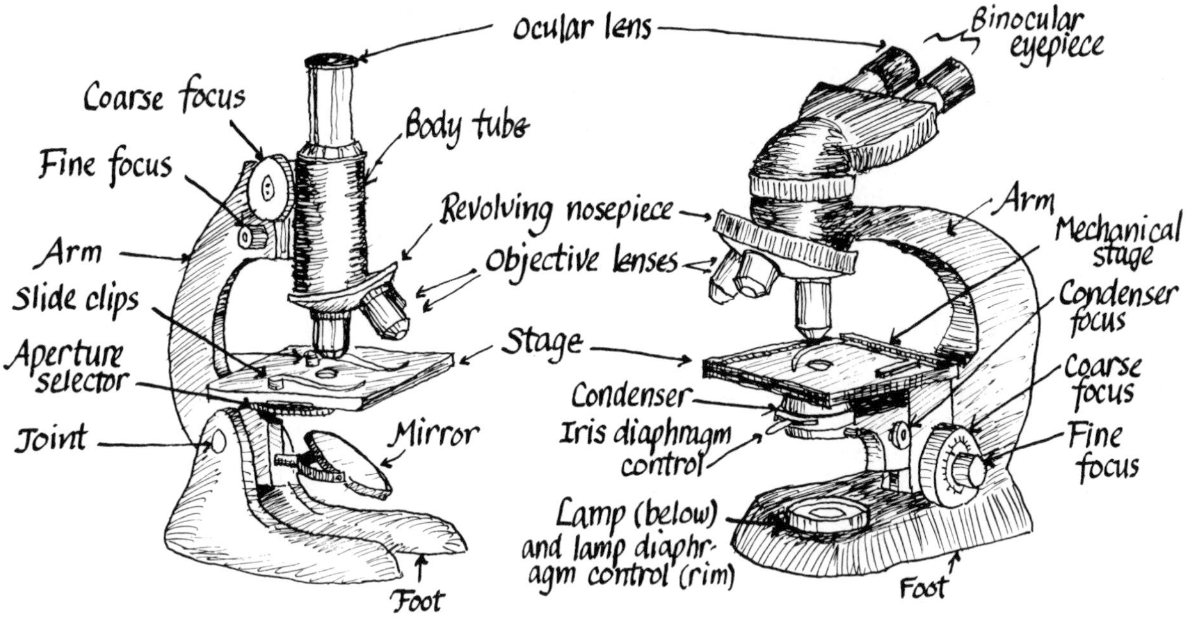

STRAIGHT INCLINED

Using the Microscope

The first requirement is *light.* The earliest microscopes used the sky, and so can yours if it has a mirror and the day is clear. But you will probably use a lamp, either a built-in one or a separate one. The second requirement is an *object* to look at; pick a well-stained section of tissue (you should see a vivid patch of color if you hold the slide up to the light). Always start with a known, easy-to-see object every time you use the scope, and never assume it is set up correctly when you get it out.

1. Simplest case: straight scope as above, no lamp or condenser. Select the lowest-power objective (the numbers on the side of each objective give the power) and rotate the nosepiece until it is in line with the body tube and clicks into place. Put your slide on the stage so that the patch of color is in the center of the hole in the stage. Now, watch *from the side* and turn the coarse focus knob to bring the objective as close as you can to the slide. It can't touch it, but with the higher powers it is possible to jam the objective right into—and even through—the slide, and this is the best way to utterly ruin a nice $100 objective. So make it a habit always to look from the side whenever you lower the objective. Now, look through the ocular—you will see only a blur. Tilt your mirror in all directions until you have the brightest blur you can get. Then use the coarse focus to raise the tube. If you're not sure which way that is, for heaven's sake *look from the side!* Once you have refreshed your memory, begin to raise it, looking through the ocular. Do this very slowly. At some point you will begin to see a colored blur, then an image. You may go right past focus so quickly you didn't notice it—in this case, lower the objective and try again. Once you have it, use your fine focus to sharpen the image. Then arrange your mirror again to get the maximum light, and turn the aperture selector to a setting that gives you an evenly lit view with no glare.
2. Intermediate: condenser, separate (simple) lamp. Focus the scope as in (1) above. Then focus the condenser in this way: take a sharp pencil and hold it just in front of the light source. Looking through the microscope, focus the *condenser* while moving the pencil point back and forth slightly across the front of the light, until the sharp shadow of the pencil appears. The condenser is now in focus on the lamp and on the slide. If you can now see the glowing filament of the lamp, change the condenser

focus just enough to make it disappear (or, better, if you can get a piece of ground glass, put it somewhere between the lamp and the condenser). Now, adjust the iris diaphragm by moving the lever that sticks out of the condenser until all glare is gone. If you stop it down too far, you will start to see dark haloes around things. When it is just right, you are lighting $^{2}/_{3}$–$^{3}/_{4}$ of the objective lens. You can see this by pulling out the ocular lens and looking down the body tube; the whole lens is illuminated on the fully open setting, and when you have reduced the bright disk to $^{2}/_{3}$–$^{3}/_{4}$ of this size, the diaphragm is set correctly. (You won't see this after you replace the ocular lens, of course.) Usually you can just set the iris diaphragm to give the best-looking image, but if you aren't sure of its setting, check it by this $^{2}/_{3}$–$^{3}/_{4}$ rule. (One of the most common causes of the cry, "I can't see anything!" is having the iris diaphragm set so far open that the specimen is completely hidden in the glare.)

3. Most advanced case: condenser, built in lamp. In this case, focus the scope as in (1). Then find the lamp diaphragm control (usually the rim of the lamp housing) and turn it so that only a pinpoint of light comes through. Now look through the scope and focus the condenser until you can see the sharp shadow of this pinhole, forming a bright many-sided spot in the center of the field. It if isn't in the center, use the centering screws (usually two long screws sticking out of the lamp housing near the diaphragm control) to center it. Then open up the lamp diaphragm until light fills the field. Finally, adjust the iris diaphragm as in (2), above.

Now you can set up any microscope properly! Take note of how it looks–particularly of the fact that *the condenser is nearly all the way up.* If a scope doesn't look like this, you can bet it isn't set up properly.

If you want to change magnifications, you will do this by selecting another objective lens. They are marked on their sides with a number indicating magnifying power as a factor—for example, 4 ×, 10 ×, 43 ×, and 100 ×. These are the most common objectives. Now you must find out one critical thing: ask your instructor, "Is this microscope parfocal (pahr-fo′-kul)?" If it is, you can change objectives easily and you should only need to adjust the fine focus. (You will also have to readjust the iris diaphragm as in (2), above.)

If your microscope is *not* parfocal, you must repeat the focusing procedure given in (1), above, *every time you change objectives.* If you do not, you had better be prepared to hear the sickening crunch of metal on glass, because in a nonparfocal microscope a setting that is in focus for one objective may be all the way through the slide for another objective. Look from the side when you change objectives, and to be safe, raise the body tube before doing so!

Using the Microscope—Comfortably

Contrary to popular belief, the microscope isn't supposed to give you a headache; if it does, you're using it wrong. Here are some pointers for comfort and better images:

Focus: There are two ways to focus a microscope; with your eyes relaxed and with them straining. To avoid the latter, focus as you look *through* the microscope, not *into* it; imagine you are looking at something far away, not inside the microscope. Reinforce this by looking up periodically and training your eyes on something outside the window or across the room. If you have to adjust too much, it means you're not looking *through* the microscope.

Binocular/monocular: If your microscope has a binocular eyepiece, be sure that it is adjusted so that the distance between the oculars is the same as the distance between your pupils. You should not feel crosseyed—and if the oculars are just a bit too close together, it will make it harder to look through the microscope. Most binocular scopes have a scale showing the actual distance beween oculars in millimeters (the average is around 65 mm)—note down the one that's right for you and set the scope to that each time you use it. If your scope is monocular try to keep both eyes open when using it. It will seem hard at first, but you will be less fatigued. A prolonged wink is very tiring!

Eye distance: Your eyes must be held at a precise distance from the ocular lenses. Too close or too far and you get distortion and eyestrain. So sit squarely in your seat, not perched on the edge or bending over, and hold your head steady.

Light: Most beginners use too much light—this tires the eyes.

Fine focus: Beginners focus the scope, then put their hands behind their backs so they won't "mess anything up." But this is wrong! The hands belong in two places: on the stage adjustment or slide, and on the fine focus control. Particularly on higher powers, only a small slice of an object is in focus at any one time; you must adjust the fine control constantly to see it all. (Besides, this will help you relax and look through the microscope.)

Using the Microscope to See Cells and Tissues

We begin with human interest—your own cells. Get a clean microscope slide and put a drop of 0.1% methylene blue stain on it. Then gently scrape the inside of your cheek with the edge of a flat toothpick. Don't scrape so hard it hurts—we don't want blood! Swirl the toothpick in the drop of stain on your slide until you see the scrapings come off; then gently lower a coverslip onto the slide over the drop, putting one edge down first and lowering the coverslip as if it were hinged at that edge. (This is to keep out air bubbles.)

Set up your microscope for the low-power objective and look around the slide. (*Note: always* "scan" the slide with *low* power. And be sure you're always looking at the area under the coverslip.) When you find the cells, switch to a higher power, say 43 ×. How do you know when you've found the right thing? It helps to know what to expect—look at the picture. But also keep in mind that you will normally find a *lot* of the right thing (dirt tends to be unique). You may see lots of bacteria, deeply stained clusters of regularly spaced rods or dots. The cells you see are thin, flat ones that line your mouth and slough off constantly: a disposable liner! Their *nuclei* are the most obvious feature, but you can also see small dots in the rest of the cell (the *cytoplasm*); these are the *mitochondria*.

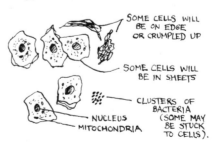

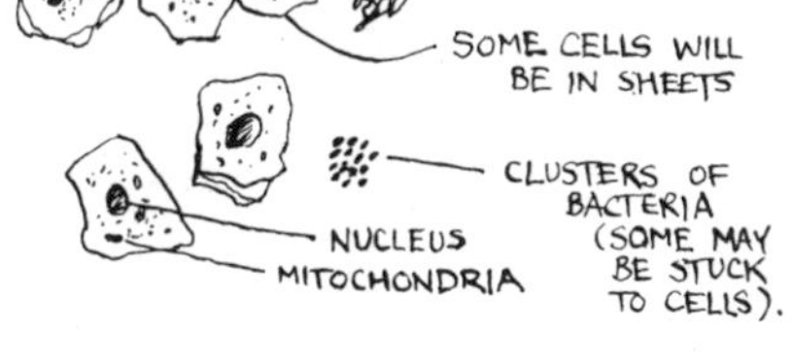

Prepared slides of organs from other mammals are also available. These are made by cutting very thin slices of preserved, hardened organs. (What do you think you'd see if you tried to look at a whole liver?) Once cut, the sections are stained with two different colors or more, most often with ones that leave nuclei blue and cytoplasm pink. Always look at the sections on low power first and try to decide how they relate to the whole organ. Here are some things to look for.

Intestine: Here the *villi* will look like fingers poking into the space in the intestine (the *lumen*). Each finger has a core of blood vessels (do you remember where they go from there?) and an outer layer made up of very special cells —long, rectangular cells in even rows, with nuclei so evenly lined up that they look like a blue band marching along the cells. At the end of the cell near the lumen, you can see a deeper pink band; this is a tuft of *microvilli*—miniature fingerlike processes that increase the cell's surface area just as the villi do for the intestine as a whole. You can also see the circular and longitudinal *muscles* that surround the intestine's inner structures. They produce the movements of peristalsis.

Artery and vein: Examine these structures while keeping in mind your observations from dissection. Which one is the artery? How can you tell, and how does this relate to your encounters with arteries and veins in dissections?

Lung: Look at this particularly carefully under low power. The *alveoli* (al-vee′-uh-lye), the tiny air sacs, are not round as usually depicted, but irregular (the lung is partially deflated). You can identify *capillaries* by the red blood cells inside them. The *bronchioles* (bronk′-ee-ohles) are of two types: those nearer to the trachea will have hairlike projections called *cilia* on the cells lining them; these beat like tiny oars to sweep mucus up out of the lungs (where it has trapped dust and such) and up to the back of the throat where it is swallowed or spat out. If you smoke, the cilia are paralyzed and the mucus stays in the lungs until you cough it up. In the bronchioles deeper in the lungs, there are fewer cilia, and finally none; but in their place are great, pale honeycomblike structures—cells specialized to produce mucus. You may even see a gob of pale blue mucus just being released by one of them. Single cells inside the air passages are *macrophages;* they engulf the particles that escape the mucus. You may see some of the dirt, also—it will look quite black. Why do you suppose the cells forming the alveolar wall are so thin?

Connective tissue: This, which is mostly in the way in dissection, is really what keeps you from falling right off your bones and running all over the rug. The star-shaped cells are *fibroblasts,* and they sit like great spiders spinning the threads of fibrous protein which keep you in shape. Around them you will see other cells, notably *fat cells,* which look like doughnuts with oversized holes (the fat is in the hole—it has been removed in the preparation of the sections), and, of course, capillaries and larger blood vessels.

Questions:

1. Answer the questions in the text above.
2. Draw and label each type of tissue. Identify all items underlined above.

The Protists

Long ago, biologists used to argue about some of the organisms they saw under the microscope: are they animals, or are they plants? The

1880

argument caught people's attention, and it is still presented in some popular books as a live issue. Since the explosion of knowledge about molecular biology, however, biologists have had far more worthwhile things to argue about. Their response to the old issue has been, "Don't ask silly questions." In fact, they have simply cut the Gordian knot and created a separate kingdom for these troublesome creatures, one in which they may flaunt their plantlike and animallike characteristics without shame. Having been victims of man's obsession with reducing everything to neat, either-or choices, these innocents are now set free to be just as peculiar and contrary as all living things are. Their kingdom is a varied and astonishing realm, and if you have learned well how to make your microscope work for you, then your passport is in order.

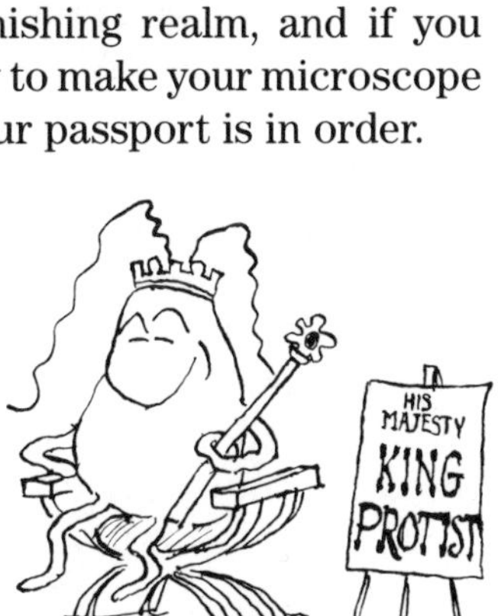

1980

Protozoa

The protists whose characteristics are, on the whole, more animallike have kept the old name *protozoa*—"first animals." The name voices a hunch that these creatures were the earliest ones to leave off photosynthesis and go in search of a free lunch. Their means of *going* have been the basis for their classification, and they fall into three neat groups (if you ignore a few nonconformists): those that row, those that use propellers, and those that use tank treads. Of course, we can't just call them that, so they are *ciliates*, *flagellates*, and *sarcodinids* (or ameboid protozoa). There are other names, but we'll leave it at that.

Paramecium is a typical ciliate. Get a drop of a culture and mix it with a goo like methyl cellulose to slow them down. (If you haven't got goo, just tear up a bit of lens paper and put the shreds into the drop of culture.) Add a coverslip (remember, one edge first) and scan your slide for slipper-shaped beasts. You should see some immobilized, or at least slowed down. Stop down your iris diaphragm (or select a smaller substage aperture) until you can see the fringe of *cilia*—do you see why paramecia may be said to "row"? These hairlike oars are all joined at their bases in a honeycomblike network, which causes a rhythmic beat. Not only do the paramecia row with cilia, they also eat with them. Add a drop of dyed yeast at the edge of the coverslip and watch what becomes of them. The *vacuoles* into which they are ultimately sucked function as temporary stomachs. After the food is digested, the inedible remains are ferried to a spot toward the blunt end of the paramecium (but away from the gullet) and dumped overboard.

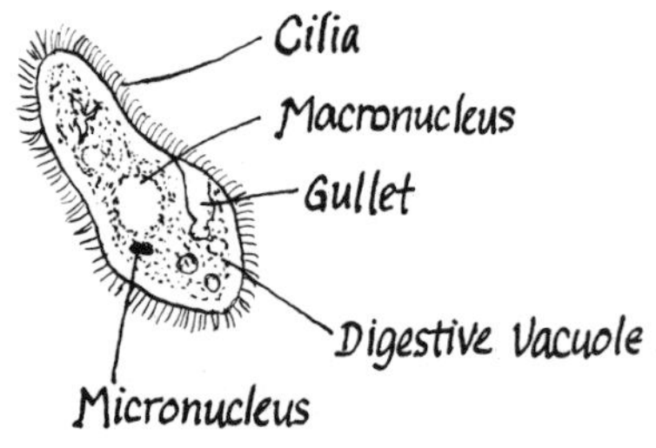

A large, more or less central, clear area is the nucleus—one of them. The other is a small dark spot nearby. The big *macronucleus* seems to contain copies of all the genes the organism is using at the moment; the *micronucleus* has a complete set of genes and can be swapped with another paramecium in one of the strange varieties of sex with which nature abounds.

You may see a sort of heartbeat if you look closely. First, a clear spot appears; then, it abruptly vanishes and, at the same moment, several clear streaks appear, radiating away from where the spot was. These vanish and the spot reappears. What is this? The name, *contractile vacuole,* doesn't help much. It is really, to continue the nautical analogy, the bilge pump. The paramecium (like you) is salty inside, but it lives in fresh water. Osmosis (see "The Cell Surface") is constantly forcing water *in,* and, to keep from blowing up, the paramecium is just as steadily pumping water back *out.* Your specimen may die (the lamp is hot), and when its pumps shut down, you will see what osmosis can do.

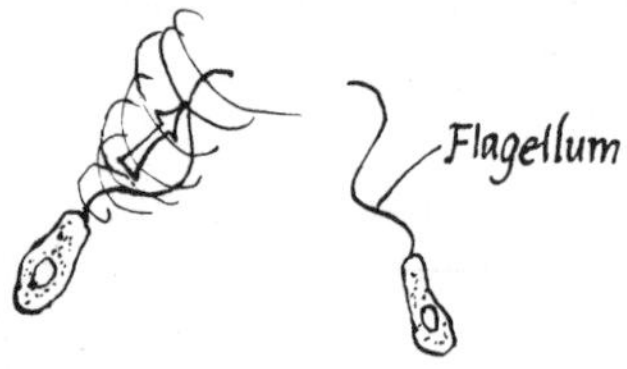

Flagellates may be represented by *Chilomonas,* or else you may simply consider the *flagellum* of a nonprotozoan, *Euglena,* which will be mentioned later. In any case, the whiplike propeller of a flagellate, the flagellum, beats rapidly to form a tornadolike vortex. The lower pressure in the vortex sucks the body of the flagellate along. Talk about pulling yourself up by your own bootstraps!

Sarcodinids, ameboid protozoa, have the strangest locomotion of all. They flow along like a blob in a science-fiction movie, using their outer membrane to stick to a surface, then rolling down new membrane from the front and pulling it up in the back, just like tank treads. If you have a culture of amebae, use a pipette to pick up material from the bottom—fine, fluffy stuff, usually—and make a slide of this. Don't use methyl cellulose, and be sure to lower the coverslip *slowly.* Then find the fluffy material under low power and look in and around it for amebae. (They may not move for a few minutes.) Can you see how the cytoplasm moves? What happens when the ameba tries to go in two directions at once? The organelles in the ameba are similar to those of the paramecium: contractile vacuole, digestive vacuoles, and so forth. If time permits, add a drop of chilomonas culture to your ameba slide and then ring the coverslip with petroleum jelly to keep the water from evaporating. Look at the slide from time to time, and you may see how the ameba uses its temporary appendages (*pseudopods*) to capture food.

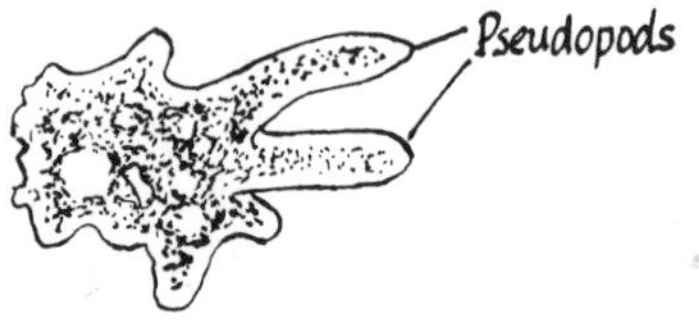

There is a fourth group of protozoa, in addition to these three. A new mode of locomotion, you ask? No; none at all. For the members of this fourth group, the *sporozoans,* are the degenerates of the family; they hitch rides (often in bloodsucking insects) and invade cells such as red blood cells. There they divide and divide and divide—until at last, with a microscopic pop, the cell explodes and releases a whole brood of protozoa. The most typical sporozoan of this type is represented by *Plasmodium,* whose four species cause the four kinds of malaria. Look at a prepared slide of a malarial blood smear. The young organisms snuggle up in red cells and look like a ring with a large stone (the nucleus) on it. As they divide, they crowd the cell more and more. Can you estimate how many are finally released from one cell?

Algae and Euglenids

There are several groups of algae, classified according to the kinds of pigments they use to trap light. They all use light like green plants, but they are single-celled. First, get a sample of a culture of *Euglena*. You won't need methyl cellulose for this. Add a coverslip to the drop of culture on the slide and scan on low power. You should easily see this pliable, undulating creature with its vibrating flagellum. Its color is derived from much the same pigments that make plants able to trap solar energy. These pigments, and the enzymes needed to exploit them, are contained in the membranes of *chloroplasts* like those of any plant. But there is also a spot of vivid red pigment, the *stigma*. This is a light-sensitive spot that allows the organism to determine where there is the most light; it is located near the base of the flagellum and has an influence on flagellar motion. Tape a strip of black paper together so that a slit of light can come through between its ends, and slip it over the jar of *Euglena* culture. After 15 minutes or so, pull it off and note the distribution of the organisms (which you can see as a green color in the water). This ability to seek out the light one needs and then swim to it is characteristic of the euglenids.

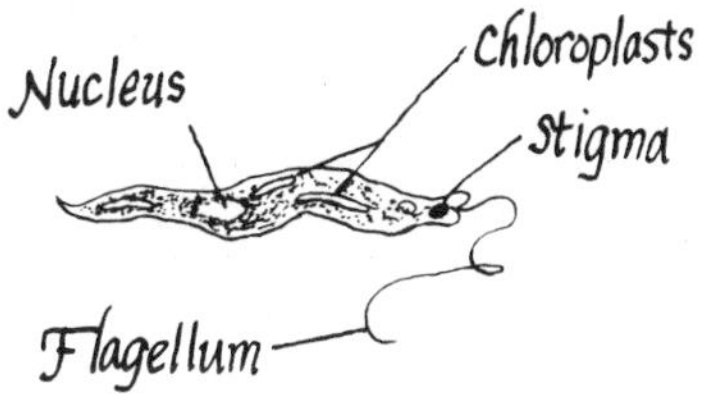

Diatoms, or "Golden Algae"

Among the algal protists is one beautiful group with golden, light-trapping pigment and glass cases—the *diatoms*. The seas are full of them, and they are also found in fresh water ponds and streams. Their elaborate shells are quite varied in form, and provide a basis for identification—also a hobby, for some people collect them and arrange them in elaborate patterns on microscope slides.

You will probably have prepared slides of diatoms, which you should examine—how many different kinds are on your slide? The patterns of spokes and such on the shells are actually rows of tiny holes which function in gas exchange. The shell is really in two parts, which fit together like a box and its lid.

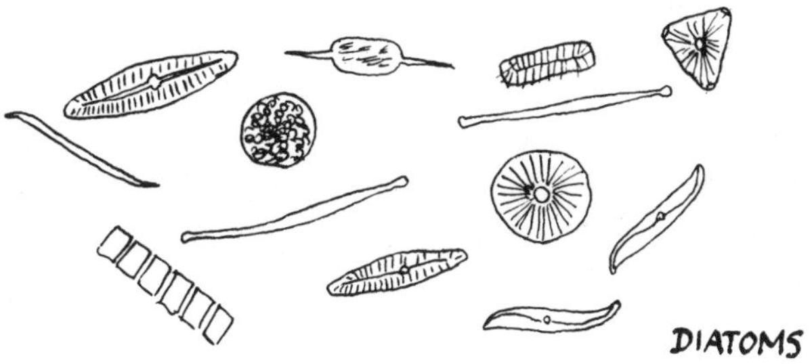

Look at samples taken from the bottom of your pond water; you should find plenty of diatoms there, gliding about. They push themselves along by releasing a detergentlike substance—chemical jet propulsion!

There may be other cultures available to you besides the ones mentioned. By all means, look at them. You aren't the kind of tourist that goes only to send back postcards, are you? There are many sights to see.

There are stalked ciliates—*vorticella* and others—that sit on a flexible stalk and use their cilia to create a vortex to suck food into their mouths. There are predators like *Didinium*, little tubs with ballerina skirts of cilia and vampire mouths, that can kill and devour much larger organisms. There are amebae that cover themselves with tiny grains of dust to make a tortoiselike shell, and others that extend long, slender rays until they look like miniature suns.

But best of all is hunting protozoa in the wild. If the weather is warm, visit a pond. Take samples from all parts, especially bottom ooze and bits of floating leaves and water plants. Anything that looks rather fuzzy is a safe bet; just avoid perfectly clear water. If it is winter, you can still find a lot of organisms at hand. Simply sample the brown material in an aquarium filter. You will find there a large number of protozoans (as well as some metazoans such as rotifers) and, if the aquarium is in a good light, algae as well. Many aquarium filters harbor a particularly good crop of amebae. You should be able to identify the major types of protozoa from your observations of their locomotion, and for further identification you may wish to go to a key to the protozoa. A large number of people pursue this as a hobby, much like birdwatching. In all your observations on protozoa, keep in mind that the microscope lamp generates heat that can boil your beasties while you watch! Don't look at a slide for too long at a time, and turn off your light whenever you are not actually looking through the microscope. For the beginner, the next most common problem is using too large a drop, so that the coverslip floats; this allows fast-moving protozoa to dive in and out of focus, an effect that can make the most experienced microscopist queasy.

Questions:

1. What kind of range of sizes of protozoa have you seen? (Put this in terms of relative size.) Which other groups of organisms do you know with a similar size range? How large, for instance, is a blue whale compared to a shrew? Can you think of factors that might influence the possible range of sizes?
2. In the ameba, did it seem to you that the cytoplasm was of an even consistency throughout, or was it thicker in some areas and thinner —more watery—in others? In terms of this thick-thin idea, describe what happens when a new pseudopod forms.

On Being a Metazoan: Cnidaria and Porifera

As you have seen, the protists can get along very well in quite a number of ways. Their major problem is one of size; being single cells, they can only get so large, and the majority of them

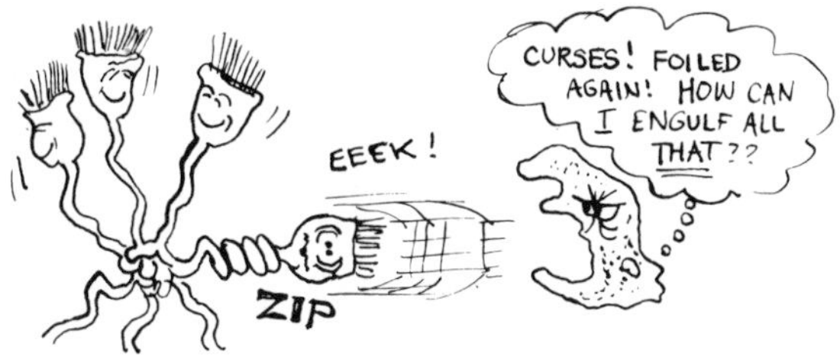

are just the right size to be eaten. One of the simplest ways to avoid being eaten is to get bigger, and probably the colonial protists like *Volvox* are doing just that. There are many such colonial forms. Up to a point, this works; but the colonies themselves suffer a number of limitations. The really important step comes when the members of a colony begin to specialize, some taking over one function, others another. To get a feeling for the difference between a colony and a group of specialized cells, think of a large family in which everyone had a separate set of pots and pans, bought groceries and prepared meals individually, maintained separate furnaces and hot-water heaters and individual bathrooms—what a mess! what waste! what chaos! Real families avoid such duplications, although they (unlike ant colonies) do not today tend to assign functions rigidly—we hope!

Metazoans and *Metaphytes* are specialized, many-celled animals and plants, respectively. We have come to the parting of the ways here, and we will consider the members of our own kingdom, the animal kingdom, first.

Most many-celled animals have a way of dividing functions that is similar from one group of animals to the next, at least in general; this is not too surprising, since they are all facing similar problems. We can broadly distinguish three types of cells and in most animals these correspond roughly to three layers of embryonic tissue also:

First, there are the outside cells—the skin or barrier enclosing the organism. These are tightly joined together (for obvious reasons!) and are therefore called *epithelial;* they may be like flat paving stones or bricks in a thick wall. Because they are the barrier between the animal and the world outside, some outside cells have *sensory* functions, either as single cells or as parts of complex organs (for example, the lens and cornea of the eye).

Second, there are inside cells, also epithelial. These are fundamentally involved in food gathering and waste removal, and they tend to have more complicated surfaces than the outer epithelia so as to have a greater area in contact

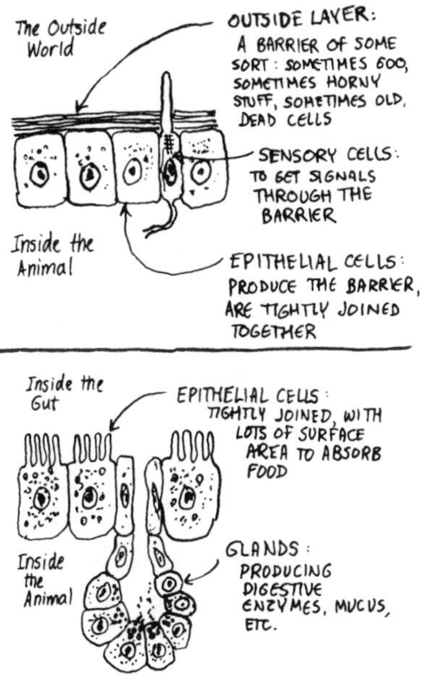

with food or waste. Some of these cells form glands which secrete digestive juices, and in many animals some of these cells form respiratory organs.

Third, there are middle cells: these are quite variable in form but usually include amebalike cells that seem to be involved in removing bacteria and repairing damage, and other cells, often star- or spindle-shaped, that produce fibers of tough material, or are themselves able to contract. Often there are other middle cells that produce masses of crystalline or jellylike material to give a shape to the animal and provide something for the contractile cells to attach to.

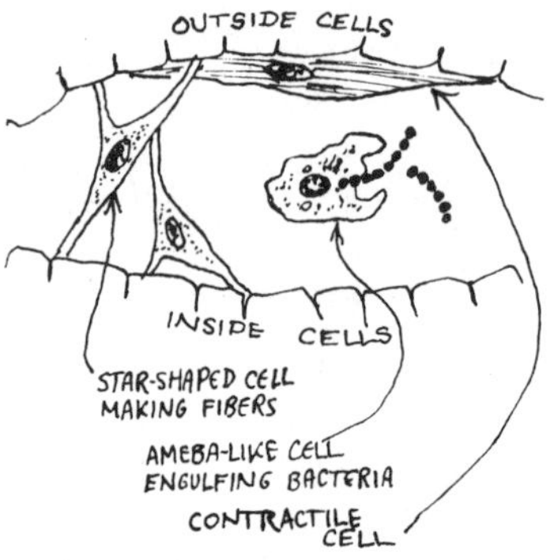

These categories are even weaker than the usual categories you encounter in courses, so don't take them too seriously. They are put here in the hope that they will help you to sort out in your mind all of the various tissues that go to make up what is called the "body plan" of the major types of animals. Let's go on to these types, or *phyla;* we will be considering nine of these that are most widespread and important.

Porifera

It isn't easy, being a sponge. Even the biologists sometimes mutter about kicking you out of the animal kingdom, and a lot of them won't even call you a metazoan but say you are a *parazoan!* The problem is that sponges have a rather loose kind of organization—run them through a meat grinder, and they just sort themselves back out again (see "The Cell Surface"). And their embryonic development breaks all the rules. They are a sort of primitive renegade that never went in for organs or nervous systems. You might say that if an anarchist were to build an animal, a sponge is what would be produced. Yet, in all the seas of the world, there are the sponges, doing quite nicely.

The casual observer might take these brightly colored, rather shapeless masses for a mold or something. But if you were small enough, and came too near one, you would find yourself tugged by a powerful current right into an opening in the sponge and thrown about inside a labyrinth whose walls are lined with whips. These whips beat ceaselessly, and at their bases lie collars of sticky meshwork in which you could be trapped like a fly in a spider's web; the mesh is reeled in from time to time, and whatever is caught in it is absorbed. The labyrinth leads to a larger chamber in which there is a hope of escape, for the current sweeps toward a hole. But when you arrive, you may find this guarded by a sieve, dooming you to bumble about the labyrinths again until you are finally caught and eaten!

This science-fiction scenario is really happening, daily, to billions of tiny marine organisms. Take a close look at the sponge specimens available to you and locate, if you can, the many tiny openings *(ostia)* and the one or more large exits *(oscula).* You will have trouble with some specimens due to their complex structure, for sponges vary in their arrangement of parts. After being duly impressed with the varied shapes and huge sizes of the prize specimens, ask your instructor for a nice, simple syconoid (syke′-uh-noyd) sponge. This is a sponge like the one in the drawing of the Grantia sponge, and it will

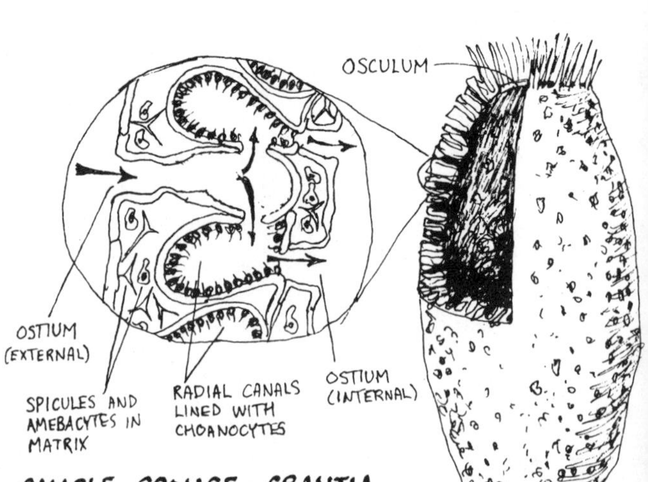

A SIMPLE SPONGE - GRANTIA

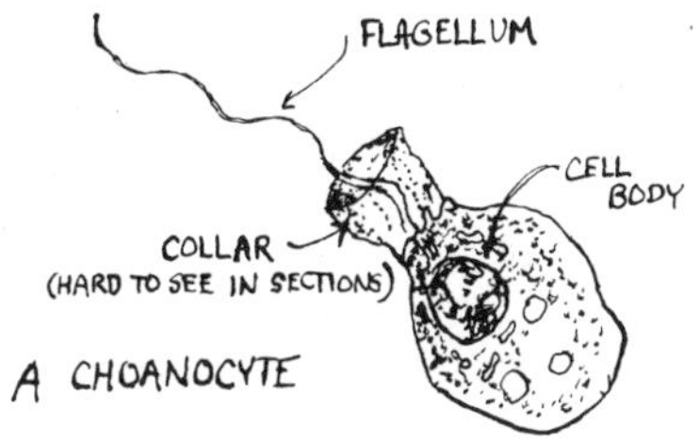

probably be available to you whole and sectioned. Study both, noticing especially the central cavity (the *spongocoel*), the bristling rim of the single osculum, and the labyrinth of radial canals in the thick walls. In the sectioned material you can see a thin outer layer of cells, the *epidermis*, a darkly stained layer of cells lining the radial canals, the *choanocytes*, and in between these a *matrix* containing *amebocytes*. The choanocytes are the cells that bear the whips (flagella, of course) and meshes in the description given earlier—the mesh being a collar around the base of the flagellum. Do these three types of cells remind you of anything? In the sponges, the amebocytes have roles in spreading nutritive materials from choanocytes to other cells and in forming temporary reproductive structures at certain times, as well as the engulfing of foreign matter.

You may see sections of other types of sponges, most of them showing a greater complexity of structure as an adaptation to greater size. You may also see slides of *gemmules*, sporelike structures formed by freshwater sponges in the fall. But be sure you see a preparation of *spicules*, either a prepared slide or a homemade preparation set up by soaking bits of preserved (not dry) sponge in dilute KOH. These wicked-looking barbs and stars are the sponge's skeleton, or part of it, crystalline harpoons made by amebocytes. They help to make sponges highly unpalatable; the sponge cannot run away, but it can be obnoxious!

Cnidaria

Here are the beasts that everyone admits to be true metazoans. They are, due to their appearance, still liable to be thought of as nonanimals, but they have a rather higher level of organization than do sponges. You will see this when you look at sections. Cnidaria (nid-air′-ee-yah) or, as they used to be called, coelenterates (suh-len′-ter-ates), are fairly well known to most people in their showier guise as corals, sea anemones, and jellyfish. But an old laboratory standby is found clinging to the under-side of leaves in ponds all summer and is easy to raise indoors—hydra. You should examine live hydra under a hand lens or a dissecting microscope, against a dark background.

The hydra that appears on the next page is hungry and undisturbed. If you have transferred yours to a slide or a watch glass recently, they will be quite round and tiny, and their tentacles will be short. Leave them quietly (and without strong light) for a few minutes, and the hungry ones will look like this.

Inspect your hydras, and sectioned hydra slides as well, to identify the structures shown here. Your hydras may have young ones budding off of the column; this is asexual reproduction, and generally goes on all summer. In the fall, gonads form as shown (except that only a few species have both kinds) and sperm are released from the testes, fertilize an egg in the ovary, and the egg is subsequently shed. (Eggs can survive the winter; adults cannot.) The *column*, you notice, is really a bag: a stomach of sorts, but blind-ended so that the "mouth" is really mouth and anus alternately. The gut lining *(gastrodermis)* consists of cells that secrete mucus and enzymes for digestion, and cells that absorb the digested food. The epidermis contains mus-

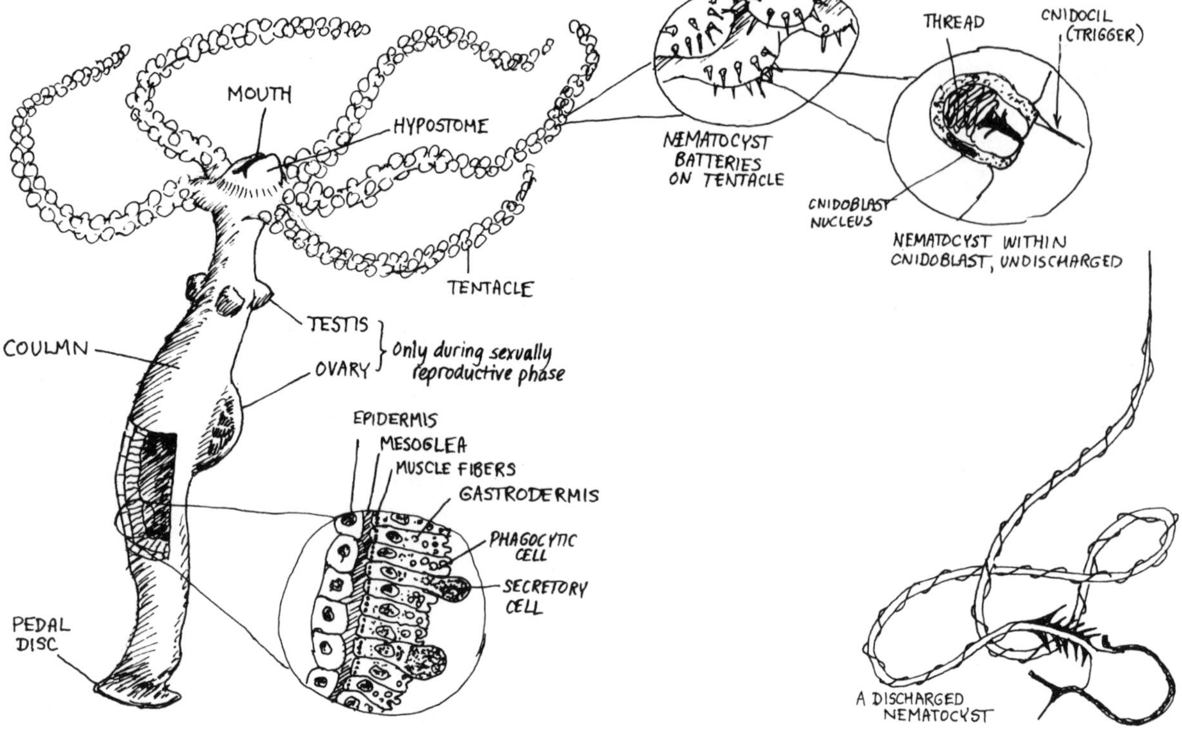

A TYPICAL HYDRA

cle fibers running from mouth to foot, and the gastrodermis contains circular muscle fibers. By filling the gastrovascular cavity with water, the hydra can become rigid and these two types of muscles can collaborate to change the animal's shape—which proves that even the spineless can accomplish something if they keep their mouths shut!

The most remarkable thing about hydra—and all other cnidaria—is the *cnidoblast*, the cell that bears the *nematocyst*. This cup-shaped cell makes a deadly little packet, which, when its spikelike trigger (the *cnidocil*) is touched, pops open and turns inside out (under explosively high water pressure). The elongated filament that results may be one that wraps around things, sticks to things, or, most commonly, one that pierces and injects a paralyzing drug into prey. These are the organs with which cnidarians capture prey (the larger jellyfish can kill small fish) and defend themselves; and the nematocysts can give even us a painful sting, in large numbers.

When your hydra are fully extended, very *gently* introduce a few *Daphnia* or well-rinsed brine shrimp into the water near their tentacles. Carefully observe what happens when the prey is captured. In this process, the hydra uses about 1/4 of its nematocysts, but these are replaced in a few days.

Place a hydra on a slide, add a drop of water and a cover slip, and focus your microscope upon one of its tentacles under high power. Locate one or two undischarged nematocysts; they look like eggs. Stop your condenser diaphragm down until you have a lot of contrast, and then have someone *gently* put a drop or less of 5% acetic acid at the edge of the coverslip. Watch closely. This can be repeated, but only with a fresh hydra. Do you think the choice of the word "explosive" above was justified?

A final note on hydra: the reactions to prey, and capture of prey, are directed by no brain at all—just a simple network of nerves covering the whole hydra. Did its behavior seem purposeful to you? Do you think it really was?

The hydra is only one type of cnidarian. Other cnidaria take the same body plan, turn it upside-down and let it float free or swim weakly; this is a *medusa,* and the familiar jellyfish is an example.* You will probably have some preserved jellyfish and perhaps a stained, embedded specimen of one like aurelia. Though the *shape* is different, the *body plan* is the same as a hydra's. The *manubrium* is an extension of the mouth, and bears nematocysts.

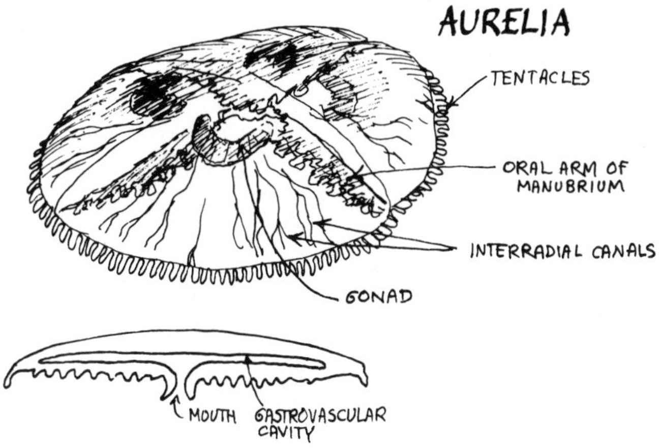

You may also have slides of colonial cnidaria like *Obelia,* with each *polyp* (the hydra form) in a shell, the *hydrotheca.* Notice here how some individuals specialize as reproductive structures *(gonophores)* and others specialize as defensive and feeding structures *(gastrozooids*—gas-tro-zo′-oids—the ones with tentacles, of course). Now stop a minute. We started out talking about the difference between *colonies* of cells and *organisms* made of specialized cells. Now we have *organisms* (made of specialized cells) forming colonies of many organisms which *themselves* become specialized . . . and that seems like a very, very good place to stop.

Questions:

1. There is a protist that seems to be a free-living choanocyte, and it has a relative that forms small colonies. Why is this of interest?

2. Give a good description of the events involved in feeding in hydra.

*Some cnidaria have both medusa and polyp (hydra) forms, and only one will reproduce sexually; so it used to be in fashion to call this "alternation of generations." But all forms are diploid and there are no spores, so today the term is reserved for plants.

The Worms That Got Organized: Platyhelminths and Aschelminths

To most people, a worm is a worm, and that's that. Aristotle agreed—when he classified living things, he put down all small nuisance-type creatures as worms (vermes), which gives us the term "vermin." We have become a little more particular since then, and we prefer our "worms" legless, but that's as far as popular precision goes. Worms have had a bad press for the most part, and most people are repelled by them. But biologists, perverse souls that they are, have been fascinated by their variety and have felt that they are terribly important. They have split them up into half a dozen phyla and claim to see all sorts of profound truths in the squirmy little things. So put aside your prejudices, and let's take a look at the world of worms to see what Aristotle missed—and the first person to say "Eeeyuchhh!" will probably be reincarnated as a robin.

Most of the interest in worms is evolutionary. But please don't get the idea that a worm is just a stockbroker who hasn't arrived yet. We haven't got any real slices of past history, and anyhow, evolution isn't a single-file march. It is more like a huge, gnarled tree. We are all out on the twigs, and we can't see the branches or the trunk; but we can guess about the number and length of the branches from the way the twigs are grouped. Some twigs seem to have come from "suckers"—branches coming right up from the lower trunk. Sponges are the extreme example for the metazoa. Other phyla have branched off at other points, and the whole sequence gives us a notion of the steps involved in reaching the more complicated forms. (By the way, complexity is equated with advancement in evolutionary language, but neither term is meant to be taken as meaning "better" or even, necessarily, "better adapted.") So don't be confused when an organism's characteristics are spoken of as if the organism stood directly in our line of ancestry; this is just a shorthand way of saying that we think that these characteristics were also found in our remote ancestors at a point just before our two lines separated. A worm that you see today has no chance of evolving into a stockbroker—or even a politician or a professor. It has its own strange path to follow now, wherever it may lead.

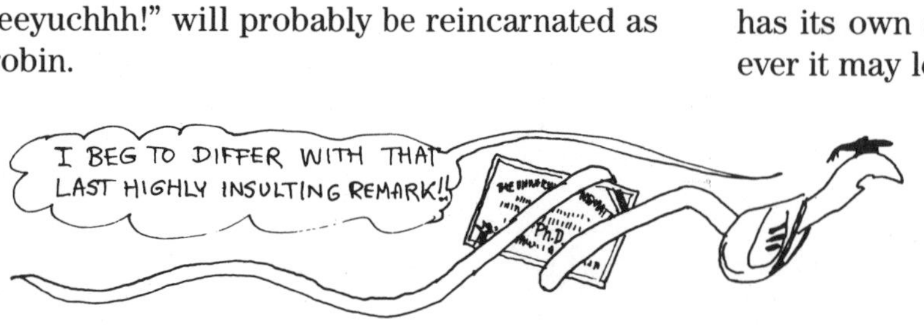

To begin with, put the earthworm out of your mind; he and his kin will be in a different exercise. The two phyla we are concerned with are flatworms and roundworms, and both of them have something we haven't seen yet—organs. Although the sponges and cnidarians had specialized *cells*, they didn't have groups of cells forming special structures that functioned as units. The cells, though specialized, acted independently. Now we add something—cooperating groups of cells, acting almost like some little organism within the organism. We call them *organs*—hence the title of the exercise. Ahem. Sorry about that.

The Platyhelminthes

The flatworms are the first group we have discussed that have true organs—for example, eyes and kidneys. The representative usually chosen is a planarian; this is a common organism to find in freshwater ponds in summer, often gliding about under the rocks at the water's edge. It is possible to fish for them using a chicken liver tied to a string.

Handle live planarians very gently; pick them up with a small brush, keep them out of strong light as much as possible, and give them a few minutes to adjust to any changes you make. First, use a dissecting microscope to observe a few moving around in a shallow container. How do they move? Are they swimming, crawling, or what? This is not easy to answer. For example, watch what happens if they bump into something. You may have enough planarians to place one in a drop of water in a depression slide and pin it down with a coverslip; look at its edges (especially as it turns over) under low power on your regular microscope. Now can you tell how it moves? The flickering band you see is a layer of cells with cilia like those of Paramecium.

These innocent-looking, cross-eyed little things are, one and all, ferocious predators. One species is so much of a gourmet that it lives exclusively on oysters, which it nibbles away at while they are still alive. Most go for smaller prey, which they trap by wrapping their thin, muscular bodies around the victim, or pinning it to a rock with a layer of slime, and then deploying that indecent-looking pharynx to begin digesting it and pumping up the broth into the planarian gut. For the pharynx is not only a muscle-bound pump but also a source of digestive enzymes.

Give your planarians (*not* the one in the depression slide; it has a headache) some hard-boiled egg yolk or scraps of fresh liver, and leave them undisturbed for some time (you may have to leave them for most of the period) in dim light. They will probably begin feeding, and you can see the pharynx at work. If you mash up a bit of a harmless dye like carmine with the food, in a few hours you can see the gut stained. You may have prepared specimens showing this. Notice the elaborate shape of the gut? Planarians are classified by the number of major branches it has; how many does yours have? But notice, also, that the gut, though branched, is still blind-ended, and waste must go out through the pharynx. The crossed eyes are simple pigment cups with associated nerve cells;

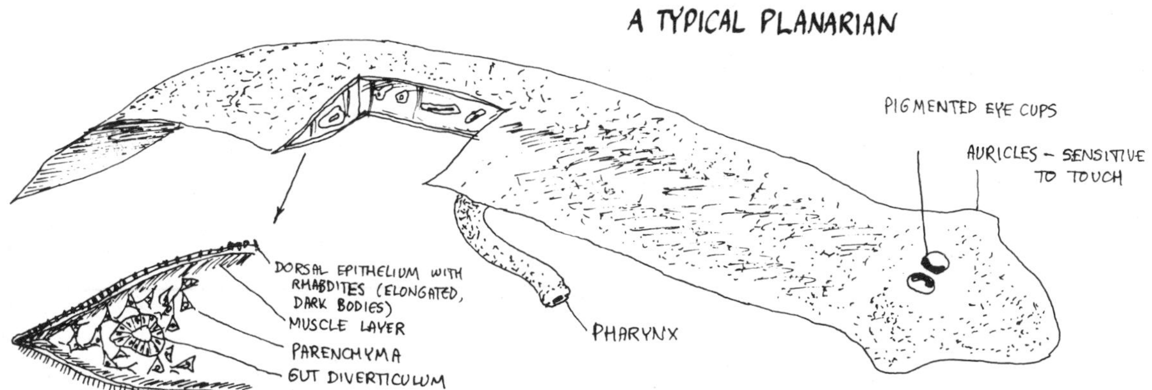

they do not form images, but they do sense the brightness and direction of light. Cover half the dish the planarians are in with dark paper and put it in the light. Where do the planarians go? Compare this to the behavior of *Euglena* in terms of preference and speed. Explain.

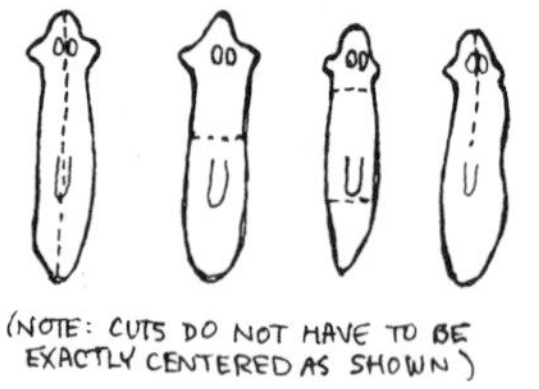

If time permits, you might try some experiments on a famous property of planarians: the power to regenerate lost parts. Transfer the planarian to the flat, wet surface of a piece of melting ice to anesthetize it, and then make one of the cuts indicated above, or others as you please. Then quickly transfer the pieces of planarian into small, clean containers of water and put them in a warm, dark place. Inspect them every day and change their water frequently (use water of the same temperature each time). Make sketches to show the progress made. It is best not to feed them, at least at first, because the danger of bacterial contamination from uneaten food is so great. In a week or two (depending on the temperature), you should see quite a bit of progress. Are some parts better at regenerating than others? Does the new part look the same as the old?

The other platyhelminths are, one and all, parasites—flukes and tapeworms. This dismal subject is postponed to an exercise on symbiosis, when you will go hunting the flukes, but you may have some specimens (safely preserved!) of such creatures now. The flukes are mainly distinguished by their large, elaborate suckers with which they hang on to your liver, gut, and such; and the tapeworms, which inhabit the gut only, are distinguished by their having sacrificed the digestive system entirely and devoting themselves to producing yards of little segments *(proglottids)*, each with a complete reproductive system and little else. When fertilized, these break off and, by a variety of disagreeable methods, get into a new host. Both flukes and tapeworms have completely lost the epidermis—their middle cell layer produces a cuticle covering. (Why? Well, wait till you have done the lab on symbiosis, and then ask.) If you can look at the stained sections (not whole mounts) of any platyhelminths, you can see that they are somehow very solid looking. They have no body cavity, and our next group of organisms does.

Question:

1. What do the regenerative abilities of planarians tell you about the amount, or kind, of tissues required for life? (Compare the regeneration with the fate of a human similarly cut up.)

The Aschelminthes

So what is a body cavity? *Not* the inside of the gut—that's connected with the outside world. The body cavity is the space in which all your organs sit. It is the space—or the potential space—that makes you flexible enough to touch your toes and expandible enough to get pregnant. Why are flatworms flat? Well, for one thing, try to imagine something with solid in-

sides that was thick and round bending double! And the tapeworm can't swell up with eggs to make lots of them; it resorts to making more body segments. One kind of body cavity is somewhat simpler than ours, and it is found in the aschelminths. It is called a *pseudocoelom* and has middle cells, particularly muscle, only on the outer side; our kind, the *coelom*, has muscle on both sides so that, for example, the gut is muscular and peristalsis is possible.

The aschelminths are either a phylum with about five classes or else a similar group of five phyla—nobody agrees. But two of the most important classes (phyla?) are favorite lab creatures: rotifers and nematodes. Rotifers (rote′-i-furs) are very common at all sorts of wet locations, and besides your local pond you no doubt have a few in your aquarium filter. Wherever you get them from, make a slide and scan it under low power; the rotifers are among the smallest metazoans and most are protist-sized, but you can easily identify them by their *corona*, or crown of cilia, which often appear as rotating wheels (hence the name). The kind you are most likely to find is the bdelloid rotifer. You will notice first the whirling cilia of the corona (which can be retracted if the animal is startled); these sweep food into the mouth. The "heartbeat" you see is not a heart but a gizzard, the *mastax*, whose ridged and horny plates crush the food. The *pedal gland* produces an adhesive by which the *toes* attach to a surface—but, above all, please note the *anus*. No, celebrate it! For this lowly orifice appears here as a new thing, and its effects are wonderful. You see, the rotifer eats all the time (notice the constant ciliary motion). This is true of many metazoans, but you can't do it without a one-way gut. So many things we take for granted! Also, note the ovaries; all the bdelloid rotifers you see have

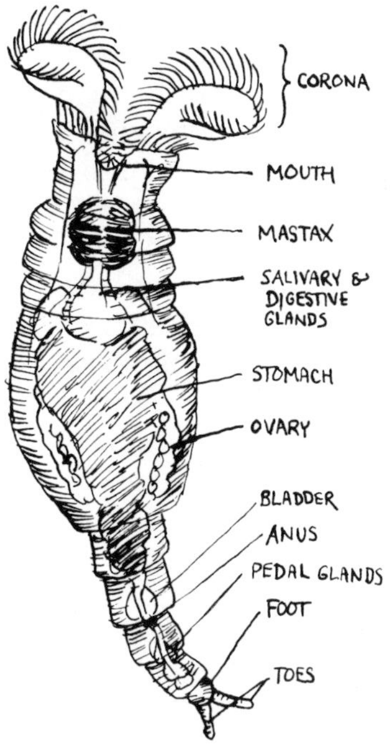

A BDELLOID ROTIFER

them. Yes, *all*. Nobody has ever found a male bdelloid! Apparently they reproduce constantly by parthenogenesis; this makes for very significant but dull genetic studies on rotifers.

Did the rotifers seem very much like worms to you? But their psuedocoelom and a few other things cause their classification with more distinctly wormy beasts—which proves either that classification is based more on anatomy and other things than on outward appearances, or else that biologists are people who don't know a worm when they see one. Maybe it proves both.

At any rate, the nematodes are wormy enough to satisfy anybody: slender and cylindrical and very pointed on the ends, and so smooth and simple on the surface that they're hard to draw. Though most are tiny, they can be many meters long. They are certainly one of the most wide-

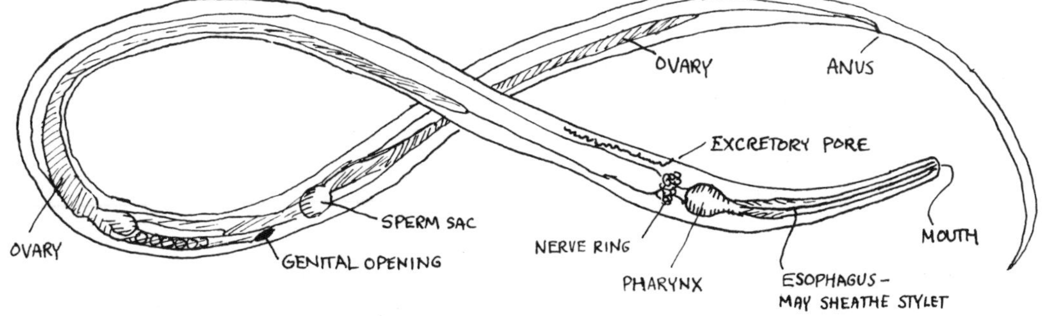

A TYPICAL NEMATODE

spread animals, living in, on, and around just about everything, living or nonliving, in the world. They live in polar ice caps and hot springs, in deep ocean trenches and on mountains, and a wheelbarrow full of ordinary dirt may contain a million of them. As every gardener knows, there are nematodes that eat plants, piercing their tissues with hypodermic-needlelike stylets and sometimes invading their roots bodily. And dog owners constantly battle these "worms," for some eat us mammals. (We are all carrying our share of nematodes, even if we are healthy.) Some nematodes eat the yeast and bacteria in homemade vinegar, and these "vinegar eels" are common lab pets today, though commercial vinegar is not a good source for them. Observe these alive, under low power of your microscope. You will easily see the bulb of the pharynx and the long intestine. The females will likely have large eggs lined up, assembly-line fashion, in their uterus. Some of them may have begun development already, for sperm is stored by the female and used to fertilize eggs at leisure. This reaches its logical extreme in the case of one parasitic nematode that appeared to produce no males. The male was finally discovered—tiny, and living out its life trapped in the female's vagina!

The nematodes are also remarkable for having a rigidly fixed number of cells—exactly 172 form the gut, and so on. (Each species has a different, but regular, number.) Imagine having a nervous system with only 200 cells! Yet these little beasts make a lot of trouble for us as parasites and crop destroyers, and probably do us good also by supplying a step in many food chains.

Question:

1. Observe the nematode's "thrashing" movements. Are these swimming movements, or not? Defend your answer. How could you test this?

The Modular Approach

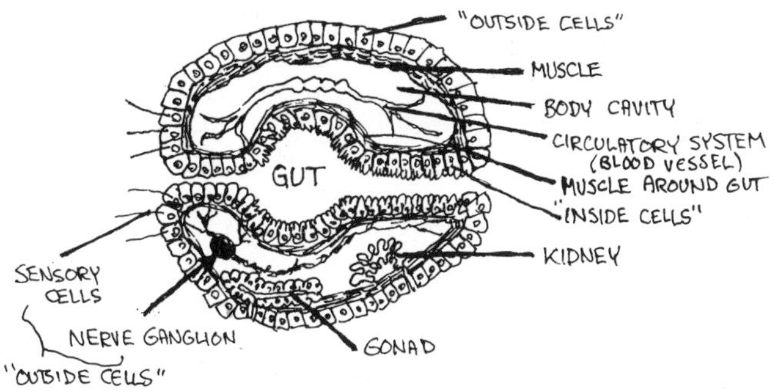

By now you must have noticed that the living world has major recurrent themes, like a symphony. Some of them are quite striking, most notably the present theme: segmentation, or *metamerism*, the presence of repeating units. You will encounter it in some of the most wildly different phyla. Does it argue a common origin? Not necessarily; it may be an often "rediscovered" theme.

The general idea runs something like this: suppose you are a pea-sized metazoan with a nice body plan made of the cell types you encountered in "On Being a Metazoan." In between breeding seasons, when your gonads shrink, you have started using the space to try out new organs. In a few thousand years you have become quite sophisticated—maybe even developing the beginnings of a brain. But who wants to sit in the mud filtering bacteria for a living and being hors d'oeuvres for every big predator? How about traveling the seas, visiting the land, maybe even becoming a predator? Size is the problem. And you can't simply make your present body plan bigger; the Scaling Law tells you that you would either suffocate, poison yourself, or starve because your volume (and hence metabolic needs) would increase faster than your surface area (and hence gas exchange, excretion, and absorption). Facing a major redesigning problem, you form a committee—and lo and behold, for the first time in history a committee solves a problem; for the committee is itself the solution.*

Each compartment, or *metamer*, does most of its own housekeeping, so the scale isn't changed, but the whole organism is now much larger. This has the advantage of escaping the need for redesigning the body plan, and it makes possible some specialization of a few segments without losing much for the organism as a whole.

In two phyla this metamerism is quite characteristic: the Annelids and the Arthropods. Annelids—segmented worms—are the closest to the fictitious example above; in particular, polychaete annelids have almost as complete a "repetition" of segments as the example. Other annelids lack so simple a form; usually the gut is specialized in the anterior segments and the vascular system also. Arthropods are usually even more specialized, with fewer segments and each one having unique characteristics; but the traces of metamerism are always there. And in chordates—look at the spinal column and its associated vessels and nerves, and consider the aortic arches!

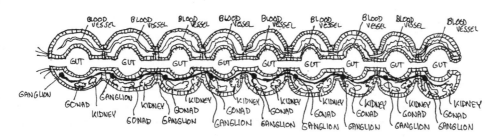

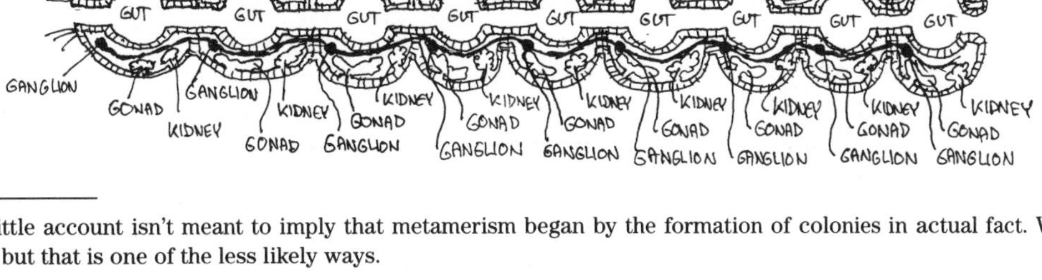

*This little account isn't meant to imply that metamerism began by the formation of colonies in actual fact. We don't know how it happened, but that is one of the less likely ways.

The Earthworm

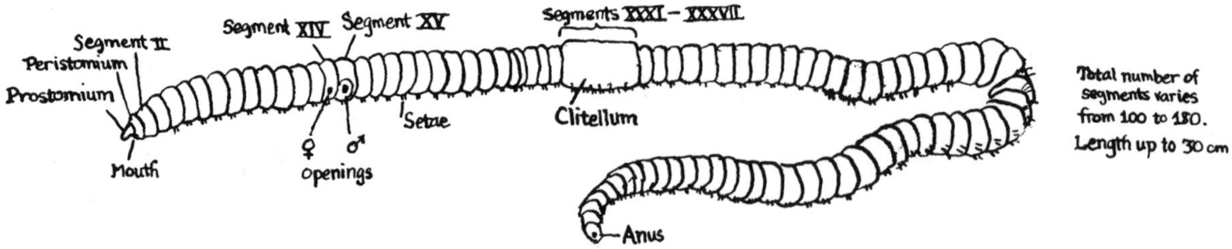

Earthworms are members of the phylum Annelida (segmented worms) and of the class Oligochaeta (worms with few *setae*, or bristles). The particular earthworm common in Europe and North America is *Lumbricus terrestris*, and this is the worm you will be using.

First, examine a living worm. Can you see the mechanism of crawling? Individual segments expand and contract in waves. In which direction do these waves pass along the body? Run a finger along the underside of the worm; which direction meets the most resistance? Use a hand lens to look for the setae responsible for this.

Note the regular appearance of the segments. Only a few specialized regions can be seen.

Be sure that you can identify the dorsal and ventral sides, and the anterior and posterior ends. (The clitellum is always nearer the anterior end.) Now anesthetize your worm in chilled water or in a chloroform jar; when it is quiet, pin it out in a dissecting pan with one pin through the prostomium and one through the last segment, with the worm's dorsal side up.

Take the pan to where you will be doing the dissection and fill it with water to cover the worm. This is essential, as many delicate structures stick together and are easily missed in a dry dissection. Now, if you have sharp scissors, insert one blade into the dorsal surface just posterior to the clitellum and cut forward. Make sure you are cutting just the body wall and not the intestine below; continue the cut to the very first segment (the peristomium). Now, working from the posterior, begin to open out the incision. You will have to cut through partitions that divide each segment from its neighbor in order to do this; use a pointed scalpel blade or your scissors, and cut the partitions as near the body wall as you can. The diagram below shows where these cuts are to be made. Do this for several segments, then stop and pin the body wall out so that the cavity is wide open. The pins must lean sharply away from the worm so that they won't be in your way later. (If you don't have sharp scissors, use a scalpel, but be *very* careful not to cut too deep.) The drawing on the next page has the large white organs (seminal vesicles) pulled down and the esophagus moved to the right.

If you examine the intestinal region of the worm, you will see most clearly that each segment has a pair of nephridia, a pair of branches of the dorsal blood vessel (you may have to scrape off the chloragogen cells to see these), and a pair of small ganglia on the ventral nerve cord. (By the way, it is characteristic of animals with a ventral nerve cord that the main ganglia at the anterior end are wrapped around the mouth. One wonders if there is a certain obsession with eating . . .) Each segment has also its own bit of intestine and a wall between itself and its neighbors.

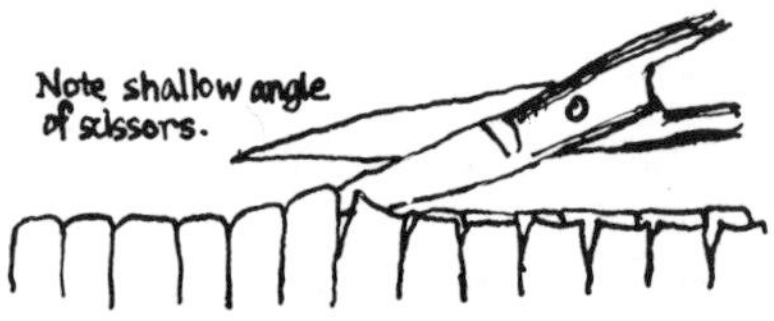

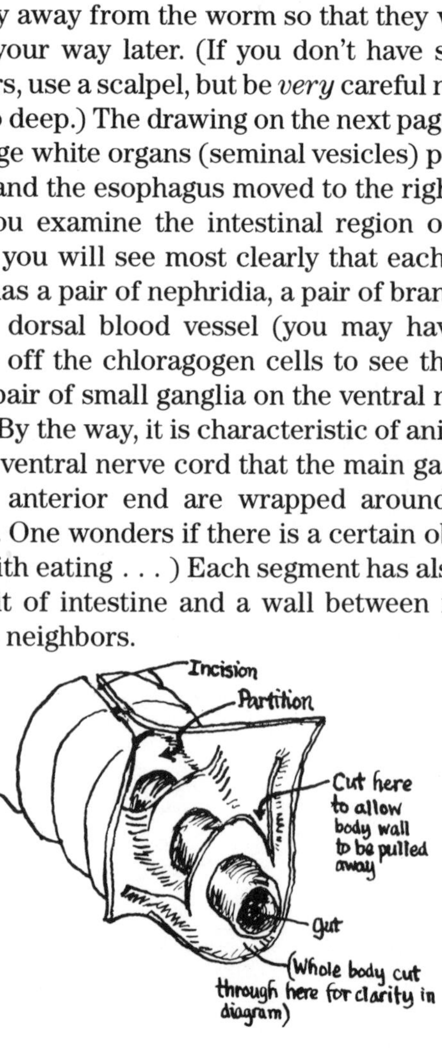

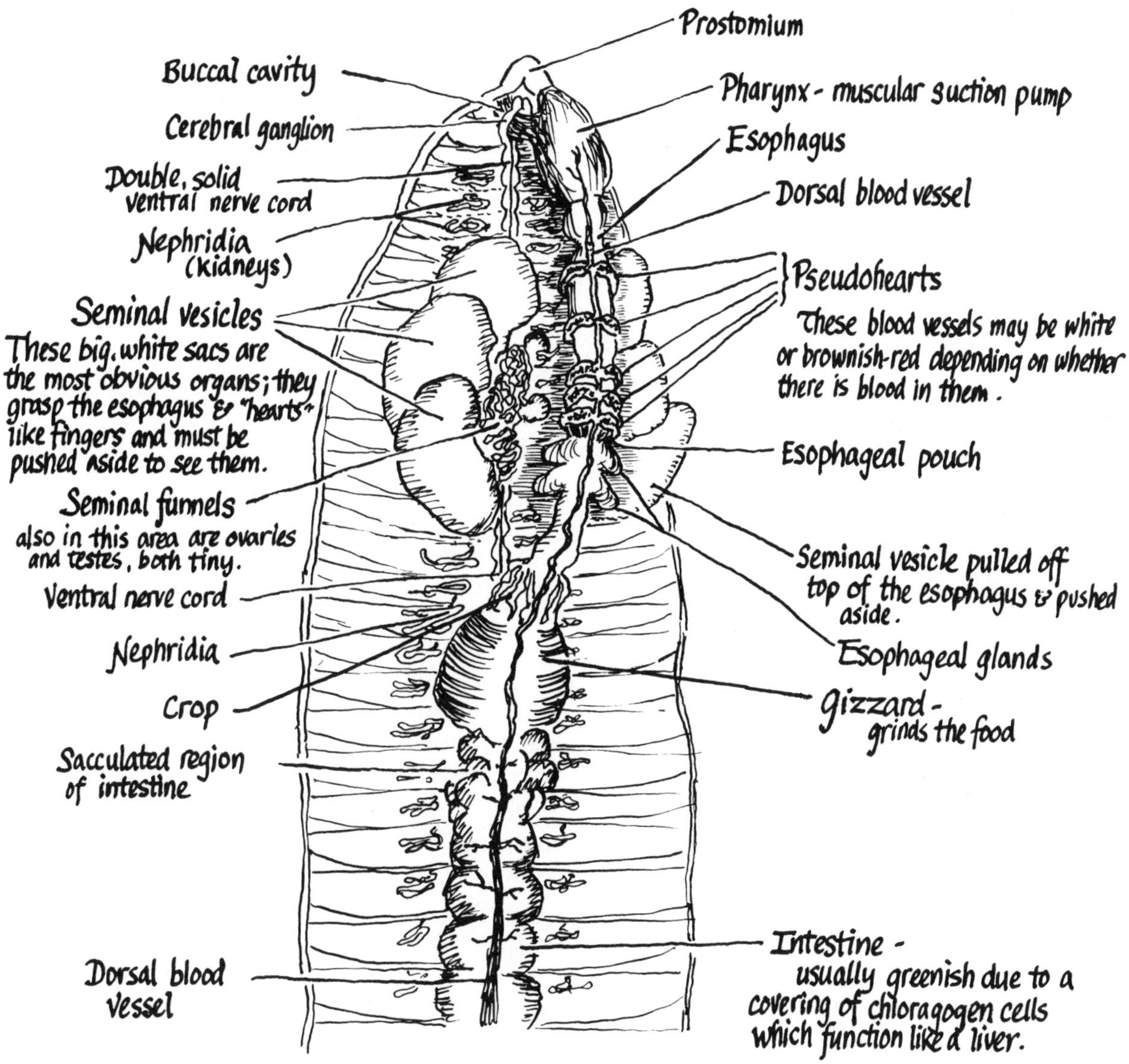

This compartmentalized plan allows annelids to get to be some of the biggest spineless things around. In Australia one type of earthworm regularly reaches 3 meters in length.

But usually you find, on close inspection, that not all of the segments are identical: those at the anterior end often have the monopoly on sex organs, and specializations of the gut as well. The tendency is called *cephalization*—head-forming. Among other annelids, it produces bizarre specializations, especially among marine forms.

The Clamworm

Clamworms are members of the phylum Annelida, also, but of the class Polychaeta, those with "many setae." The common name may derive from the fact that they, like clams, burrow in the damp sand in intertidal zones; their scientific name is *Nereis virens* or *Neanthes virens*, though what they have in common with the beautiful Nereids of mythology (except a wet habitat) is not clear. Fishermen often use them as bait.

These polychaete worms, unlike oligochaetes, are studded with appendages. Examine one with a hand lens or a dissecting scope. Some marine polychaetes have very long setae that will penetrate soft tissue and break off inside, producing pain and itching. Others have stout setae like those shown (top, right), which help them to pole themselves along the bottom or hold fast in their burrows. The flippers, or *parapodia*, are another aid to locomotion; the worm flexes its body from side to side, and the parapodia act as paddles; in walking, the worm makes a rowing motion with each parapodium, but in swimming the worm simply holds them extended and wiggles its body. Can you see any difference between anterior parapodia and posterior parapodia?

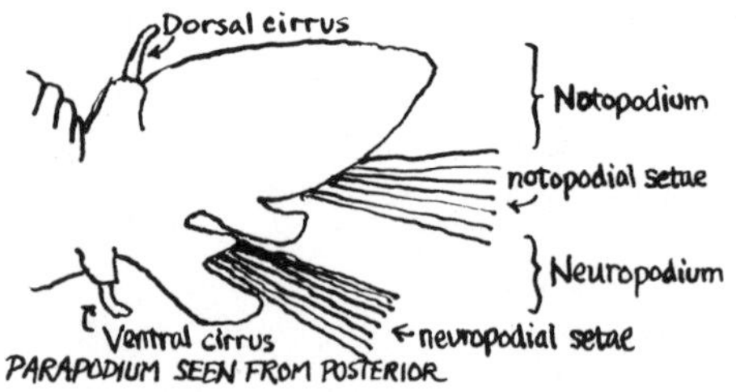

Now consider the anterior end (below): truly a face only a mother could love. Note the sensory appendages; these are typical of polychaetes. Certainly, this is an example of cephalization.

As you may have guessed, *Nereis* is a predator. Those jack-in-the-box jaws snap out (under hydraulic pressure from body contraction) and clamp on a smaller invertebrate; then a set of long, slow muscles begins hauling pharynx, jaws, and prey back in again.

Contrary to what you might expect, *Nereis* lacks specialized gonad-bearing segments. In fact, it lacks gonads! In breeding season, loose clumps of gametes form free in the coelom and, when full, segments simply pop open to release them. Thus, the internal anatomy of *Nereis* resembles the posterior segments of the earthworm; there is very little internal specialization.

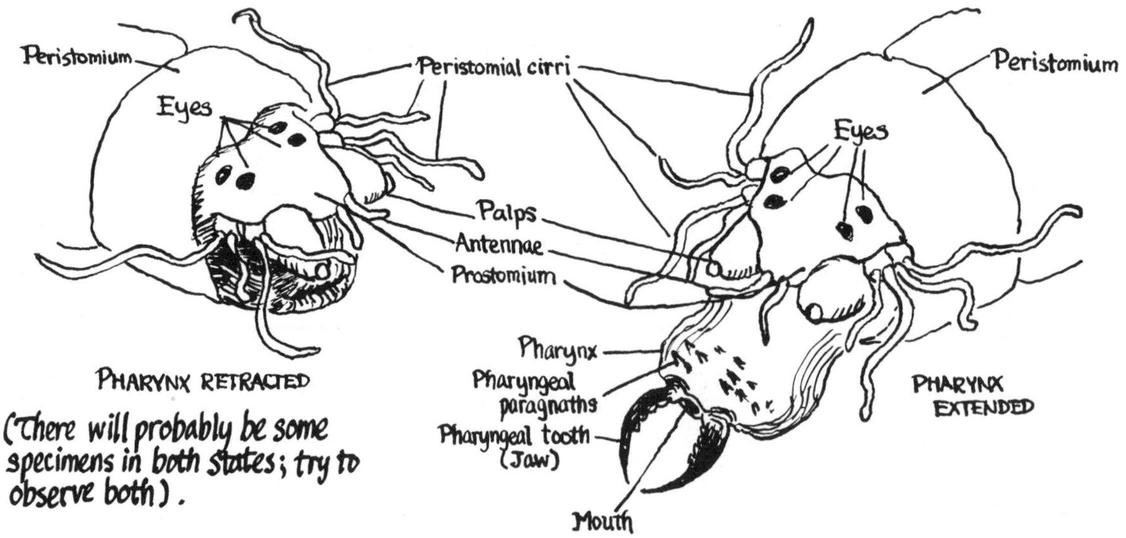

If time permits, make a *shallow* incision about 1/3 of the way from the anterior end and use a small pipette to sample the coelomic fluid. Make a slide of the fluid and examine it under high power. Dark field or phase contrast microscopy may be helpful but not necessary. The eggs are enormous, round cells; sperm are small, flagellated ones. If the animal was in the breeding season, there will be masses of gametes.

Questions:

1. Make drawings of all you have seen. Include notes on color, size, texture, etc. Answer any questions in the directions.
2. What do you consider the most striking specialization of the earthworm? Of *Nereis*? Do you see anything in their respective habitats, food sources, etc., that makes these particular specializations appropriate? Where can each afford a lack of specialization? Why?
3. What do you suppose puts an upper limit on size increase by metamerism?

The Armored Ones: Arthropods One

OF COURSE, ARMOR DOES HAVE SOME DISADVANTAGES...

How do you measure success? It's not easy, but one way might be this: if you saw an awful lot of people all driving the same kind of car, you might wonder about it. If you knew that a lot of them had been driving that kind for a long time, you might ask whether it was a good design. And if people and businesses with very different needs—retailers and messengers, commuters and car-rental agencies, maybe even race-car drivers—all used the same kind of car, you would probably agree that it must have something going for it! (Now, don't get excited. There isn't any such car!) But in the living world it is just the same, and there is one design, very different from our own, which is used by almost a million species with huge populations of individuals, has been around at least since the Cambrian era, before there were vertebrates, and has been used in every imaginable way. So, what is the great design?

Well, one part of it is armor. Hard, complete armor. Some primal beast sweated minerals one day and found it made her too hard to bite. The idea caught on. An *exoskeleton*—tough layers of a gristly stuff called *chitin*, sometimes with minerals like those in your bones, hard calcium salts—forms a complete suit of armor on these creatures, and since they need joints in it for movement, it gives them their name: Arthropods, "jointed-foot ones." This armor can be beautiful, in the gay scarlets and whites of some spider crabs or in the subtle iridescence of a beetle. (The former is real pigment; the latter, the prismlike effect of striations in the cuticle.)

The remote ancestor seems to have been the trilobite, whose backs had an especially heavy plate of armor; they lasted a long time, from the Cambrian to the Paleozoic. Though most were small, there were some the size of cafeteria trays. Their new chitin armor was probably able to keep them off quite a few menus in those far-off days, and their remote—well, let's say *cousins*, the crabs, present some problems for brainy humans to eat: ever watch an uninstructed one tackle a steamed crab?

But as the cartoon of the knight implies, there is a price to pay for living in body armor. You lose contact with the world; you are isolated as well as protected. What's more, you can't grow! Arthropods have solved this pair of problems

HEE HEE HEE HEE HEE

simply. They are covered with bristles, or *setae*, which look like hairs or whiskers but are really mechanical sensors that can detect the footfalls of prey or the stirrings of air or water. They are also studded with pits with *olfactory* (smell) *receptors* and have complex antennae, which combine many functions. And to grow, they shed the tough armor and for a time are covered with softer, stretchier cuticle, which lets them grow until it, too, hardens. Naturally, these *molts* are risky times, and molting arthropods are universally shy and retiring. Nobody wants to become a soft-shelled-crab sandwich!

If you look hard at the trilobite, you might be reminded of something. Does it look, well, a little repetitious? Yes—metamerism! In many ways, the arthropods resemble the annelids. In modern arthropods it is a bit less obvious, for there has been a strong tendency to fuse the segments that bear legs, so as to keep down the inclination a flexible body has to waggle back and forth when the legs are striding fast. This fusion produces a leg-bearing *thorax* (not to be confused with a mammal's thorax) and even a fused head and thorax *(cephalothorax)* in some arthropods. An extra shield of chitin called a *carapace* may protect these structures. (The last two terms are sef′-ah-lo-thor′-acks and care′-ah-pus.) There is also a tendency to alter the appendages to serve a variety of purposes, so that no two are just alike. (Some sets will remind you of one of those do-everything pocket knives.)

The Chelicerata

Besides the extinct trilobites, there are two major types (subphyla) of arthropods. The chelicerates have pincerlike or fanglike appendages above the mouth, and their heads and thoraxes are fused. They lack antennae.

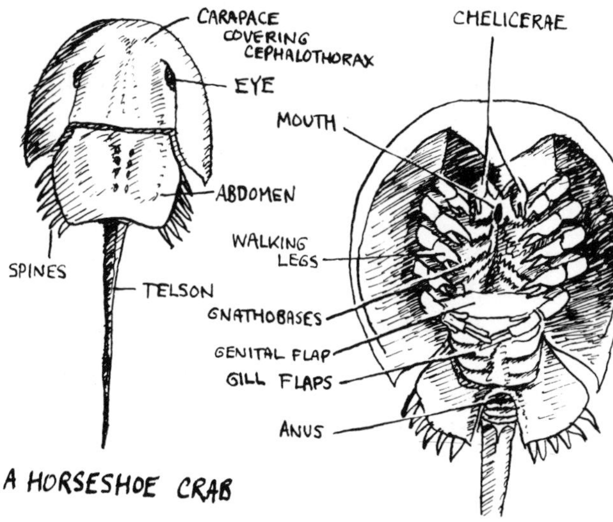

A HORSESHOE CRAB

Examine a horseshoe crab. This fearsome-looking beast is really a mild-mannered scavenger, and the wicked-looking telson is used as a lever, not a weapon. You should have no trouble seeing how far the fusion of segments has gone in these animals. Compare the legs; are they varied in form? The base of the leg has a rough surface, which helps grind up food before the chelicerae pop it in the mouth; this is the *gnathobase.* As we will see often in this phylum, legs have many uses! By the way, the larvae of horseshoe crabs bear a strong resemblance to trilobites.

Spiders are also chelicerates, adapted to land and a predatory life. They're best known for their spinning "silk" (not really the same stuff as silkworms' silk) for webs and other purposes, and for their fanglike chelicerae, which inject a nerve poison into their prey. The victim is then liquified by regurgitated digestive juices, and the resulting rich broth sucked up by the spider—a liquid diet. The four pairs of eyes on most spiders are not able to form sharp images, but jumping spiders, which pounce on their prey, can do so. You can easily recognize these by their relatively heavy cephalothorax and legs, and if you move your finger around in front of one, it will carefully follow the motion.

Mites and ticks are also chelicerates, and in these, fusion has reached the limit—their bodies are all one piece. Most are blind, and heartless as well (literally); many are parasites, biting others of us for skin or blood meals—fortunately, most will eat and run. Examine all the specimens available, and note especially their common chelicerate features; compare their degree of appendage variation with that of other chelicerates.

The Mandibulata

Other arthropods have mandibles (chewing mouth parts), antennae, and generally less fusion of segments. One class of mandibulate arthropods is the class Crustacea. They are famous for their complicated, branched appendages, their calcified exoskeleton, and their compound eyes—or perhaps more for their flavor, for they include crabs, shrimp, and lobsters! Our specimen is not so delicious, though widely eaten (when not preserved for dissection): the crayfish.

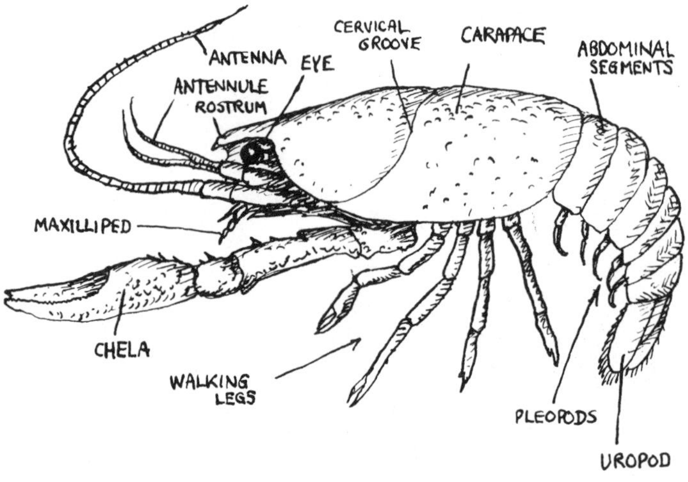

Begin by examining the external structures as shown. (For clarity, the only ones shown are those on the left side of the body.) How many appendages has this beast got, anyhow? You will have to be careful in your count, for there are four pairs of mouth appendages alone! They include the mandibles and three maxillipeds, of which the third is largest; these last are leglike food-holders. Most crustaceans are very versatile feeders—scavengers, filter feeders, predators, or herbivores, as opportunity permits. So, it is not surprising that their mouth parts are involved-looking. By the way, the pleopods are sometimes called "swimmerets," but they are used for reproductive purposes more often—crustacea don't swim all that much when they are as big as this. The uropod (together with the telson in the middle) is a powerful tail fin, however, and the big muscle that fills the abdomen can snap it forward to allow quick escapes.

Now cut away the side of the carapace on one side as shown. You will notice that the edge of the carapace is free; you are exposing a chamber that is always open to water. The feathery things inside are gills. Wiggle one of the legs on this side back and forth. What happens to the gills? Why? What good would this do?

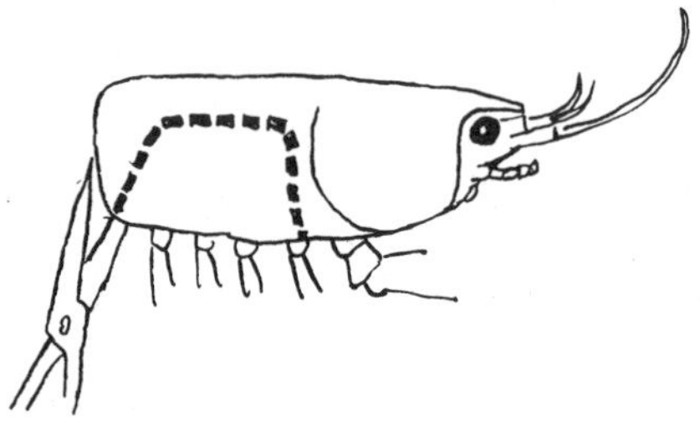

Cut off a bit of one gill and examine it under a hand lens or dissecting microscope. Remember, to increase gas exchange an organism must increase surface area.

Now put the crayfish on its ventral side and cut away a strip from the middle of the carapace (below, left). Don't cut too deeply, or jab with the scissors. (Are you beginning to feel like a frustrated predator?) Lift off the flap, cutting any attachments. You should see the organs shown below, but they may be a bit messy-looking; most invertebrate organs are. Immersing your specimen in water and looking at it that way will help a great deal. Learn to identify

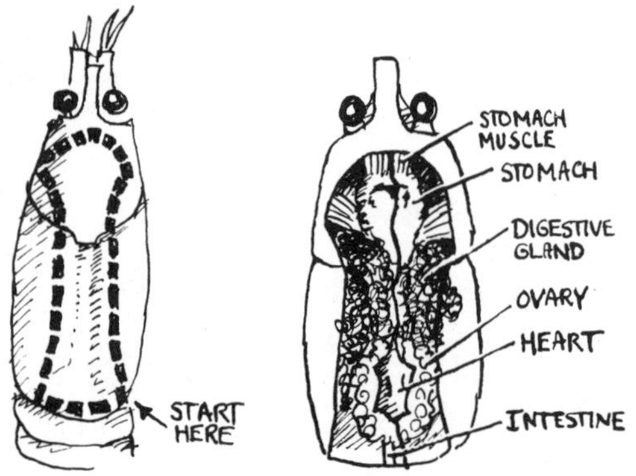

muscle, for some of it may be in the way. The stomach will be easy to locate; it is very far forward. It is thin-walled and pale, and has muscles attached to it. Cut it open, wash it out, and look at the teeth inside. This is sometimes called the "gastric mill," for obvious reasons. Large muscles beside the stomach work the mandibles—can you verify this by pulling on them while watching the mouth? Posteriorly, the gut dives below the digestive gland, or *hepatopancreas*, which not only produces the digestive enzymes but also stores the nutrients. Does the name "hepatopancreas" seem justified? (Remember, "hepato-" = liver.) If your animal is female, you may have a mass of round eggs visible in the ovary; if male, slender testes. In either case, above the gonads lies the heart, a six-sided sac with slitlike, valved openings in its sides. As in most crustaceans, the blood is a colorless to pale blue fluid bathing all the organs in the body cavity; the heart sops it up through the openings in its sides (the *ostia*) and squirts it to the head and extremities, from which it trickles back to the body cavity. Cut away the stomach, and below it notice the pair of ventral nerve cords passing around the esophagus and joining in the brain, to which the optic nerves also go. At each side of the brain is a large, thin sac attached to a flat knob—these are the green glands and their bladders, which excrete urine, the ammonia waste from metabolism. In the fresh-water crayfish, they also act like the vertebrate kidney in maintaining salt balance. But what a place to put them!

Question:

1. Answer the questions in the text.

The First to Fly: Arthropods Two

Our definition of success in the previous exercise is best met by another class of mandibulate arthropods, the class Insecta. What, you ask, do we want with another whole exercise about arthropods? For one thing, insects generally get separate billing and some texts simply avoid them, directing the reader to books on entomology. Why? For one thing, insects are by far the largest group of animals, both in terms of species and of number of individuals. Numbers and textbooks don't impress you? (Shame!) Just go into your back yard on a nice summer afternoon. The bugs are eating more of your vegetable garden than you ever will, and when you spray them with that insecticide that killed them last year, they laugh at you and go on chewing. A mob of bloodsucking types dines on your mammalian blood, and in trying to swat them away you blunder into a big anthill and set off a sophisticated defense system whose coordination makes the Pentagon look like an amateur. You beat a retreat to your house only to find moths in your closet, fleas on your dog, and a big, fat cockroach emerging from the bag of groceries. Oh, well—let the cockroach go. It will probably survive a nuclear war and inherit the planet, anyway.

JURASSIC AIRWAYS FLIGHT 601 CLEARED FOR LANDING . . .

All of this is to say that insects are the biological success story *par excellence*. This makes them a nuisance to us, of course, and often a danger (though there are beneficial types), and always a bit humiliating. But they aren't dull. They have such a range of lifestyles and forms that it can be bewildering. And the corporate "intelligence" of social insects is as fascinating to modelers of artificial intelligence as the flight of dragonflies is to aircraft designers. The big development in insects is *flight;* they were the first living things to fly. We still group them into the *pterygota*, those descended from the early fliers (even though some have now lost their wings) and the *apterygota*, those descended from early nonfliers (even though some have later acquired wings). Insect wings are not, as are bird wings, specialized legs, nor are they jointed. They are more like oars, "rowed" from inside the thorax by muscles that snap the exoskeleton into two shapes—one that forces the wings up, another that forces them down. With this system, there are a few insects, including a moth, that can pass you when you are driving at 30 mph. Others go in for hovering or rapid changes in direction. But any style of flight opens up huge possibilities in evolution, and insects have capitalized on the new options most effectively. One simple technique: hit and run. Gangs of marauding locusts eat everything in sight, and then, instead of sitting and starving, they fly to new territory. Locusts, in fact, are just migratory grasshoppers. This brings us to our subject, for the typical representative of the insects for dissection purposes is one kind or another of grasshopper, these being plentiful and large. Think about it: for much of the human race, for most of its history, the innocent-looking thing in the pan before you has been, in its devouring millions, The Enemy.

Examine your grasshopper, and identify the structures above. Different varieties may have longer forewings, and in any case, you may have to raise the forewing to see the tympanum. Also, look for three wartlike ocelli between the compound eyes.

The thoracic segments are not fused; the prominent shield, or *pronotum,* is part of the first segment. It may be smaller in your specimen. Each thoracic segment bears a pair of legs—hence, the six-legged state of insects. The wings are also attached to two segments (the hindwings are underneath the forewings). Which segment has no wings? There are fossils of dragonfly-type insects with wings on all three segments, but this has not lasted. Today, two pair is the maximum, and often one pair is modified to make wing covers (as in beetles) or balancing organs (as in flies).

Examine the legs. The long part near the body is the *femur,* and the other long part is the *tibia;* the complicated part at the end is the *tarsus* (mostly). As you know, grasshoppers are good jumpers, and the hind (metathoracic) legs show it; the size difference is obvious. But there is more than that; can you see how the tarsus is modified? What do you suppose this modification is good for?

The abdomen is perforated at the sides by *spiracles,* which connect to a system of air tubes, or tracheae. The mouth parts bear inspection; use fine forceps or a needle to pick them apart. The maxillae bear leglike *palps,* and the mandibles are chewing devices; the prominent *labrum* covers the front like a lip. The *cerci,* at the other end, are acute wind and vibration sensors—rear guards.

Now examine the wings, especially the hindwings. Use a lens if you have one. The leaflike pattern of veins you see is not decoration; the veins' thickened cuticle acts as a support, and blood is pumped through the channel.

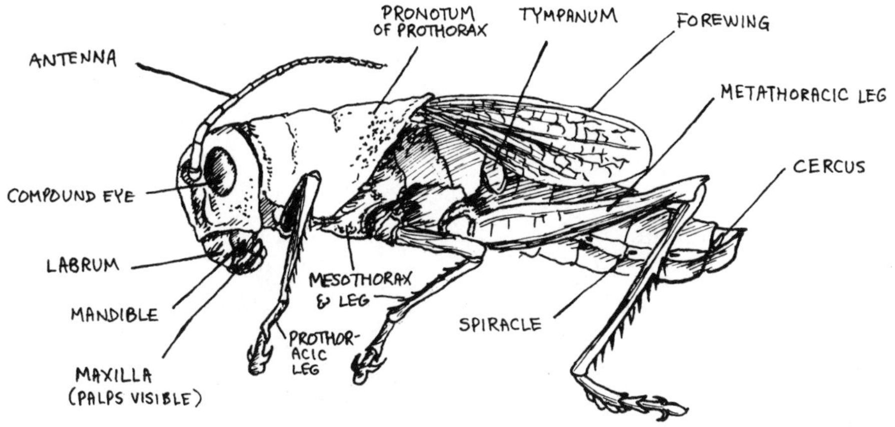

A TYPICAL GRASSHOPPER- ROMALEA

Note: only appendages on left side shown.

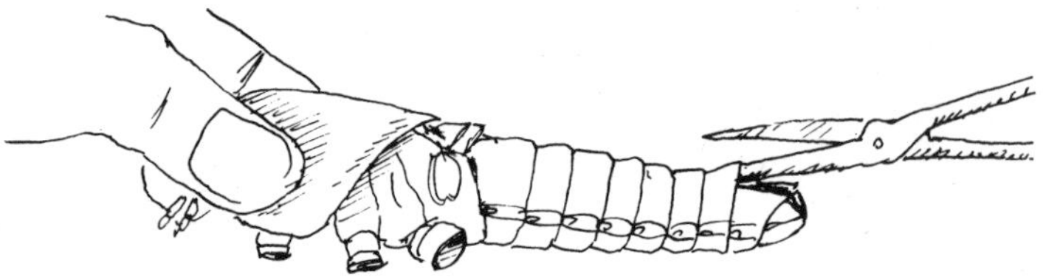

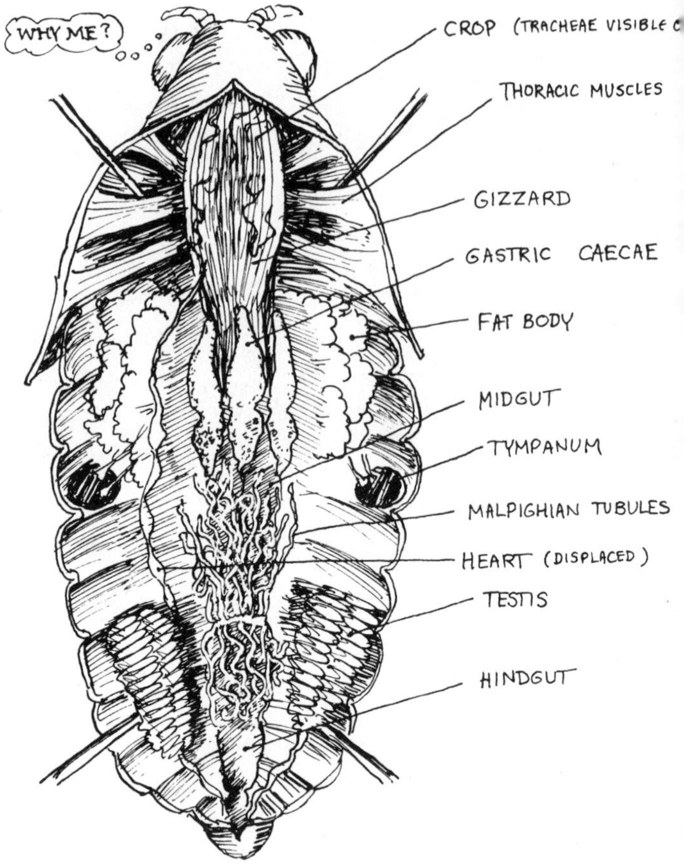

Fill your dissecting pan with water, get a sharp, fine pair of scissors, and begin the dissection by removing all legs and wings—they would be in the way. Then put the point of the scissors just under the cuticle on the dorsal side of the last abdominal segment and cut the cuticle all along the back, as shown above. Go all the way to the head. Then pin open the body wall in your pan. You may need to cut some muscle attachments to open the body fully, and you will have to pick away bits of fat and membrane, but the results should be as shown on the right.

The most obvious part (at first glance, almost the *only* part) is the gut. This should come as no surprise; the adult insect is typically an eating and reproducing machine—though sometimes only the latter.

The crop and gizzard are usually not clearly separated, and the gizzard is not as important to the grinding of food as is the crayfish's gastric mill, though it often bears teeth or ridges. Most prominent are the fingerlike *caecae*, which are attached to the anterior end of the midgut (stomach). These are pouches connected to the digestive tract; the crustacean hepatopancreas is a modification of such a pouch. Here, they seem to serve mainly as quiet backwaters where symbiotic bacteria can grow. The food, in most insects, has by this point been surrounded by a cuticle like the exoskeleton, and passes on as a sort of capsule, leaky enough to let nutrients be absorbed but hard and smooth enough to protect the gut.

Attached behind the stomach is a tangle of long, pale strands—the *malpighian* tubules, the insect kidney. While the green gland dribbles ammonia down the crayfish's face (not so bad in a watery environment), the thoroughly land-adapted insects have kidneys that dump the less irritating waste uric acid directly into the hindgut.

Farther forward, attached to the body wall or gut or both, are irregular masses of stuff of a rather loose consistency. These are fat bodies—storage sites for nutrients. Do you see any reason for their location? You may have destroyed the heart in the incision, but you may see a fragile tube running along the dorsal side of the abdomen; this heart has an enlargement in each segment—really a string of hearts.

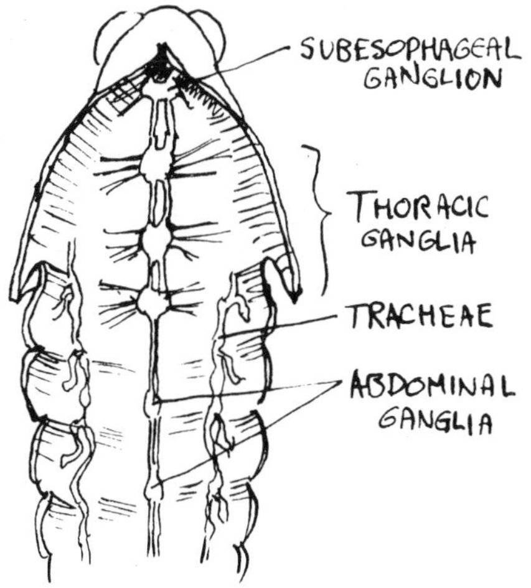

Besides eating, the grasshopper's purpose in life is reproducing. The drawings have shown male grasshoppers, but you may have a female; externally, her abdomen will have an almost pincerlike ovipositor, and internally, her gonads will be much larger than those of the male, and eggs will be visible, stacked like pancakes, within them. Dry-land life requires greater care with fertilization, and direct sperm transfer by copulation is the rule among most insects. The male usually has not only a penis but also various complicated grasping organs beside it, with which he securely fastens himself to the female for the duration. This often results in his being dragged about by the larger female, but it assures that the sperm stay moist and alive.

Now clear away the gut and fat bodies and peel off muscle stips until you can see the pale, lumpy fibers of the ventral nerve cords. In the thorax, the cords are easily seen as paired, and they are joined at three large *ganglia*—centers of complex nerve contacts permitting reflexes and such. One ganglion, you notice, serves each segment. Which is largest? (They are not drawn to scale here!) How can you account for this size difference?

At the point where the esophagus emerges is the *subsophageal ganglion*—a good portion of the grasshopper's brain. How does it compare with the thoracic ganglia? What does this tell you about the degree of centralization in grasshopper nervous systems? In the light of insect success, is this necessarily a bad thing? You may also be able to see pale, delicate tubes—the *tracheae*, air tubes. Make a slide of a bit of one and examine it under low power to see the citicular rings that keep it open despite internal pressure changes. Compare to the pig's trachea.

Questions:

1. Answer the questions in the text.
2. Compare the appendages of the grasshopper to those of the crayfish. How does the general plan differ? How much variation is there in that plan?

Happy as a Clam: The Mollusks

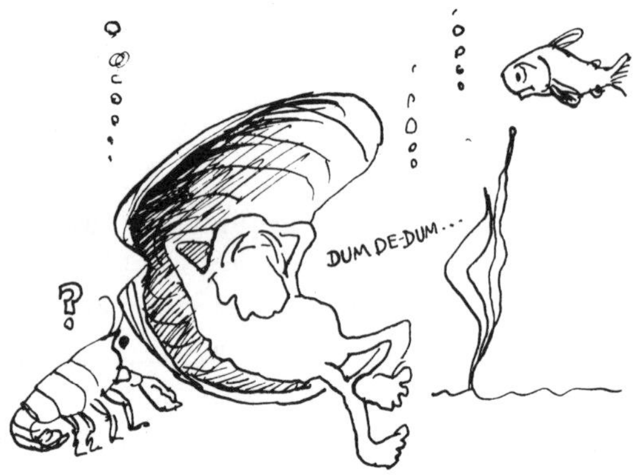

Most of the phyla so far have been sufficiently distinctive, and their members sufficiently similar to one another, that you wouldn't have too much trouble identifying the phylum to which an animal belongs. Of course, we can't leave it that simple! The mollusks are second only to arthropods in the number of species, and second to none in the differences between classes! We will talk first about some fairly lowly mollusks, *bivalves.* These little, stationary filter feeders are so vegetable-like that people use the expression "happy as a clam" to suggest mindless bliss. On the other hand, we will also consider cephalopods (sef′-uh-luh-podz), which are the largest animals in any invertebrate phylum and are lively, well-coordinated, and predatory. But both classes have a distinctive kind of feeding and digestive arrangement, and a very similar embryological development. On these and other grounds, they are one phylum. As you study these forms, you will see how widely organisms having the same basic body plan may differ as they adapt to differing lifestyles.

Bivalves

A major bit of mollusk strategy involves an alternative to arthropod armor—a shield. The animal is not encased entirely in an exoskeleton but, rather, has a distinct, shieldlike shell. This is a hard structure made of calcium salts and protein, secreted by a special flap of skin called the *mantle.* (The mantle will secrete a similar material around a foreign object, creating, in a few years, a valuable pearl.) Clams, oysters, mussels, and such, all bivalves, have two shells with a tough hinge that tends to spring the shells open, and strong muscles (adductor muscles) that hold them closed. That is why clams gape open when steamed and shouldn't when raw (supposedly alive). Take a clam and look at it: where is its head? That is not a silly question, because it does have one. The hinge on the shell is on the dorsal side, and oldest part of the shell, the *umbo,* usually points anteriorly. Most clams bury themselves anterior end down—their heads in the sand, as it were.

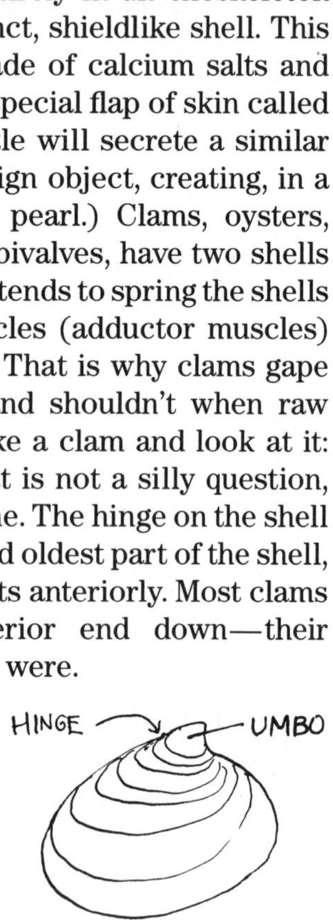

Getting into a clam can be a lot of work. Get a sharp scalpel, preferably with a number 22 blade, and use it to cut through the adductor muscles and hinge ligament—cutting against your dissecting pan, not your hand, for the blade is likely to slip.

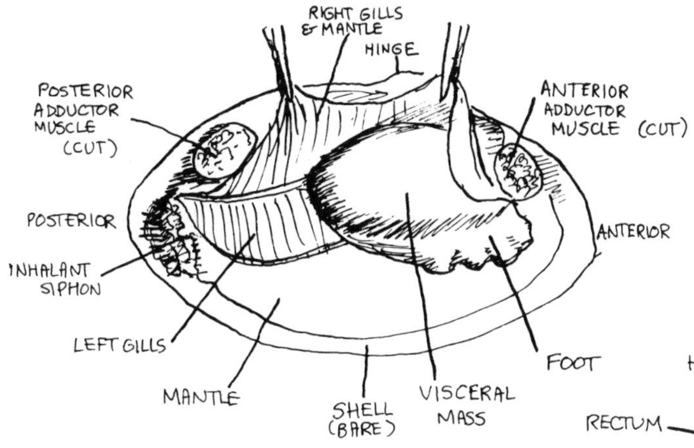

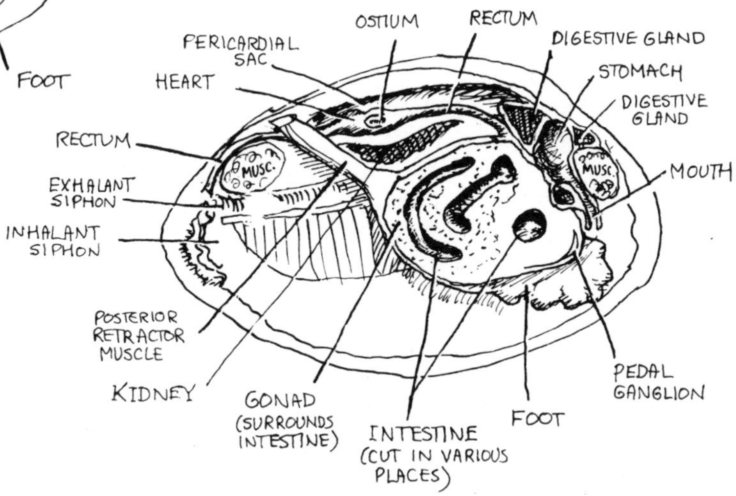

You will have to do some prying as well as cutting, and you may break the shell. Try not to cut *between* the muscles and the hinge. Eventually you will expose some gills, probably torn; fold these back or tear them off, and then get your bearings, using the drawing on the left, above (you could have its mirror image). The most prominent structure is the fat, round visceral mass. Leave the gills under it alone for now. Notice how the thin mantle covers much of the shell; posteriorly, it is thickened and complicated to form one side of a *siphon* through which water is drawn. The gills are covered with cilia, as is the mantle, and a maze of water currents sweeps over their surfaces. Most bivalves have a lot more gill surface than they need for respiration; these bivalves use the gills also to trap and sort food particles. The foot is used to burrow in the sand or mud.

Cut the visceral mass in half, holding your scalpel parallel to the shell; note that this is making a sagittal section. (Can you see this glob as a bilateral animal yet? Keep trying.) You may do a neater job by shaving off slices until you get to the middle.

You will see (above, right) that the visceral mass is solid; the only cavity is the inside of the looped intestine. Most of the solid mass is gonad, but some is muscle. Lying between the visceral mass and the anterior adducter mucle, and continuing dorsally, is the *stomach*—an empty sac—and, surrounding it and probably visible in several places, the *digestive gland*, usually green and soft. The body cavity (you were perhaps wondering if there were one?) is reduced to the space dorsal to the visceral mass; open it up to find a tube running through it.

This tube is a very strange thing: it is the clam's heart—with the clam's rectum running smack through the middle of it. You can see a thick ventricle and an earlike ostium; there are two atria also. Why the rectum-heart association, you ask? Perhaps it has to do with reclaiming something from the gut, for the heart acts as a pressurized filter for the blood, and the filtrate passes for the pericardial sac (the reduced body cavity) into the kidney, which lies just below it. But it does seem odd! If you cut cleanly through the visceral mass, you may see a whitish strip just above the foot, one of three pairs of ganglia which coordinate the clam.

Now, lift up the visceral mass by its ventral edge and, looking underneath it, cut away as much as you can without damaging the gills. You now have an unobstructed view of these food-gathering organs.

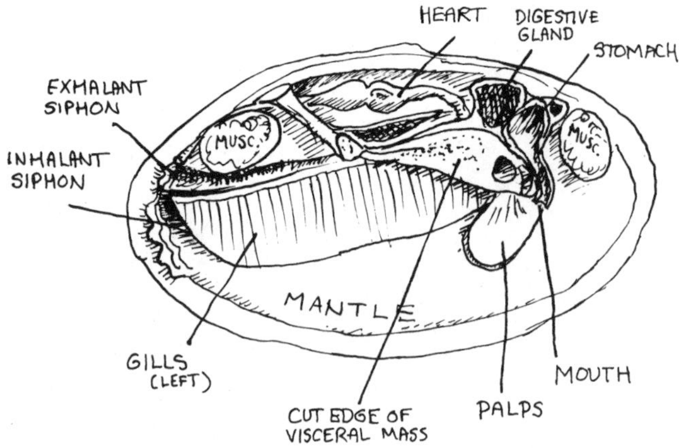

Cilia sweep water in the inhalant siphon and across the gill surfaces. There, a combination of ciliary cross-currents and gravity results in sand being dropped on the mantle (which then sweeps it out with ciliary currents) and potential food particles moving forward in food grooves along the ventral edge of the gill. The *palps* carry out finer sorting, so that, in the end, hardly anything but food (mostly algae) gets into the mouth. Examine the inner surface of the palps to see the fine ridges of ciliated tissue. On a freshly opened live clam, it is possible to drop bits of cork near the inhalant siphon and watch them be shunted around by this elaborate system.

An oddity characteristic of mollusks is the *style*, or, more properly, the crystalline style. This is a rod of protein with digestive enzymes attached, which is produced by a sac attached to the stomach and ground constantly against a hard plate in the stomach, winding up a thread of mucus bearing food particles from the mouth, and liberating enzymes. It may disappear between tides, when the animal is not feeding, but if you have time, take a look in your clam's stomach for it. Frequently, the style sac is beside the intestine, and the style can be found by probing into the intestine near the stomach.

There are other classes of mollusks which are livelier than bivalves, such as the gastropods (snails). These move about on the foot and have a characteristic coiled shell. They also have a striking molluscan innovation—a sort of tongue built on the plan of a belt sander, the *radula.* With this they go around giving rasping licks to the surface of rocks and feeding on the loosened algae. Still sound quite tame enough? There's more:

The Cephalopods

The legendary sea monsters of most cultures have included a devil-fish with huge, staring eyes and powerful tentacles. It is common because it is real—the giant squid. A sixty-foot-long predator, it has huge, acute eyes and jet propulsion. The squid's close relatives, the octopods, are well known for their learning ability and manipulative skills—one capative at a large aquarium had a habit of escaping from its latched tank to steal fish from other tanks. But these are mollusks too!

They differ in having reduced shells, tentacles, and a more centralized nervous system—in fact, a brain. A somewhat smaller squid than the ones referred to above will be the subject for dissection. As you dissect, keep looking for similarities to the clam.

Take a good look at the squid's outside, first. The directional labels are not a mistake; the tentacles are derived from the foot, so dorsal-ventral directions go the long way. Imagine a snail with its shell uncoiled and sticking straight up, to get the idea. The short muscular *arms* have suckers all along their inner sides; these suction cups have hard rims—toothed, in some squid—which leave neat circular scars on animals they grapple with (sperm whales sometimes show up with saucer-sized scars). The *tentacles* have suckers only on their flat ends; they dart out to snatch fish and pull them in until the ring of arms can grip them. The *siphon* (cf. exhalant siphon in the clam) can be aimed in any direction and makes the squid a fast swimmer; the body is rigid and bears stabilizing fins.

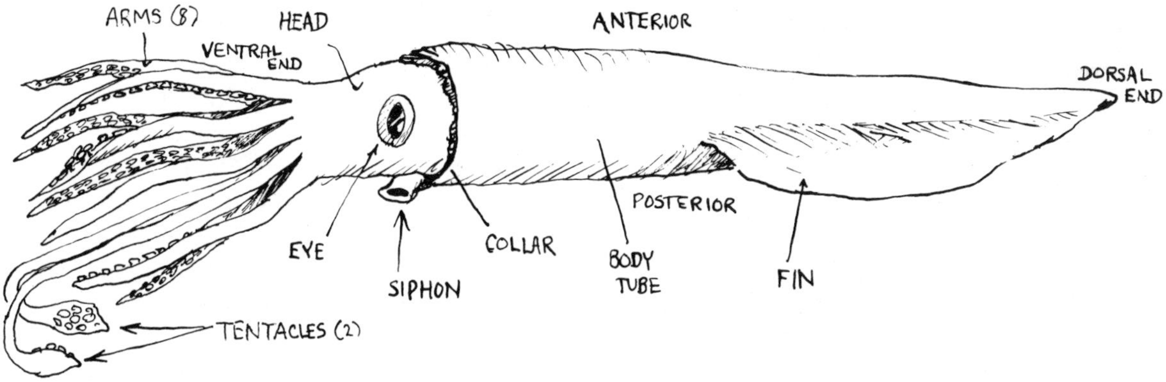

Place the squid, siphon up, in the dissecting pan and cut open the body tube from the collar to the dorsal end. Pin open the tube and identify the parts shown below.

Note the iridescent ink sac; this produces a highly concentrated solution of melanin (the brown-black pigment in human skin also is melanin), which can be squirted out via the siphon to create a smoke screen or a squid-shaped dummy while the squid escapes from an attacker. Note also the extent of the mantle cavity and the size of the gills. Here you can see the circulatory connections to the gills quite well. Also note the paired posterior venae cavae, probably quite distended in your specimen.

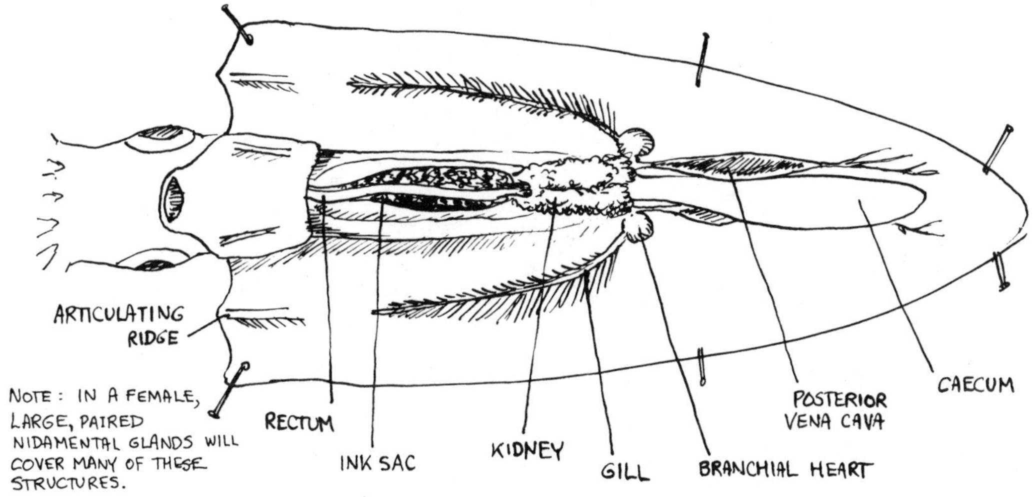

Now cut the rectum and ink sac at the base of the siphon. If you want to make ink, by the way, save the contents of the ink sac and add a tiny pinch of gum arabic to thicken it—this was a fine ink in the ancient world. Pull the rectum to one side. Lift the kidney and note, below it, the systemic heart. You will have to cut the aortae coming from the heart before you can lift it, also; then you can identify the stomach and caecum. (Their size may vary from that shown.) Down below the ink sac, observe the large digestive gland, or *liver.* Finally, cut through the head and tentacles to expose the tough *buccal bulb*—a surprise awaits you. The squid has a beak! If you can cut the mouth open farther, the rough, tonguelike radula can be seen.

Just between the eyes, you will see the large white ganglia that form part of a highly centralized nervous system. This is a smart, swift predator. If your squid is a male, you will find, below the caecum, a large testis quite separate from the vas deferens (the sperm are sucked in by cilia in the vas), and a convoluted gland that puts out packets of sperm—spermatophores. The penis is misnamed—it never touches the female. Instead, one of the two posterior arms delivers a sticky packet of sperm from the penis to the female's mantle cavity, gluing it on near the oviducts. (In some octopods, the arm breaks off and stays with the female!)

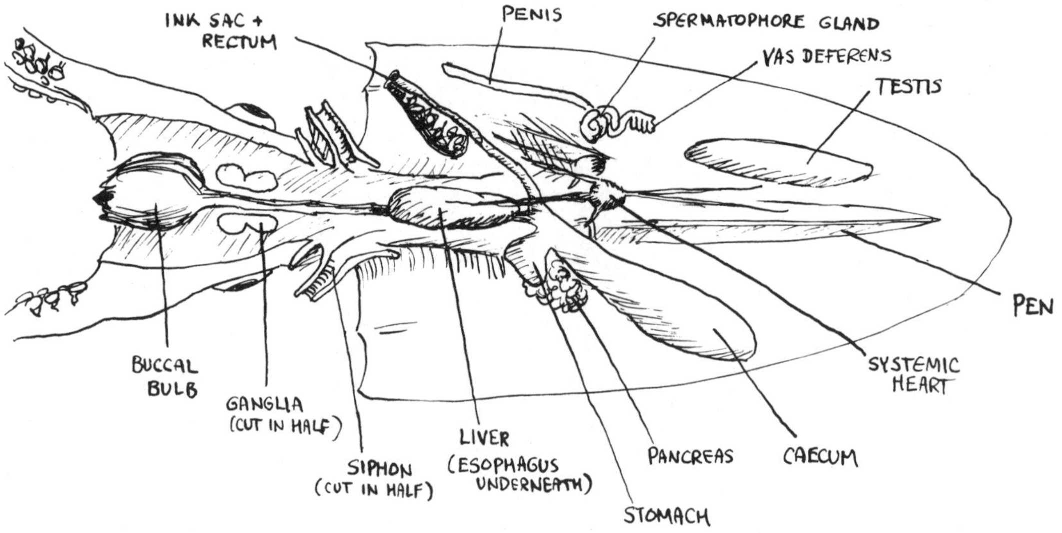

Questions:

1. Squid and clams have adapted a common body plan for two very different lifestyles—one a sedentary filter feeder, the other a very active predator. Comment on their most significant differences in this light.
2. Why has the squid got such a complex circulatory system, compared to the clam?

Strange Cousins: The Echinoderms

It is easy to say, "All life is interrelated," and burble sweet nothings about songbirds and fawns. But just step outside our phylum and see what you meet first! Fast, big-eyed, brainy cephalopods? Sensitive, social insects? No . . . take a deep breath and shake hands with a starfish. Aha—your pride protests? This is perhaps a mistake? On the contrary, the evidence is rather compelling.

The way the embryo develops is a major characteristic of living things, and it tends to be constant from one species to another, as if it were to fragile a process to mess with. Any innovations tend to be tacked on at the end of the process. Now, an early event of development, when the embryo is just a hollow ball of cells, is the formation of the gut, which begins at one end and pokes through to the other. There are obviously two ways of doing this. The *rest* of the animal kingdom begins at the mouth; we chordates begin at—ahem!—the other end. This switches around the whole orientation of the embryo (and, some pessimists would say, has a lasting effect on our outlook). The only other phylum that joins us in this back-door approach (excepting the acorn worms) is the phylum Echinodermata, which contains such creatures as sand dollars, sea cucumbers, sea urchins, and starfish. Other facets, too, of their development match ours, so it is hard to escape the conclusion that we have more in common (in terms of ancestors) with these bizarre beasts than with any others.

The Starfish

The classic lab echinoderm is a starfish. Usually thought of as dried decorations for a beach house or a seafood restaurant, these star-shaped predators are of great economic importance (they eat a lot of seafood themselves) and show the typical echinoderm structure, which is unlike anything you have seen before.

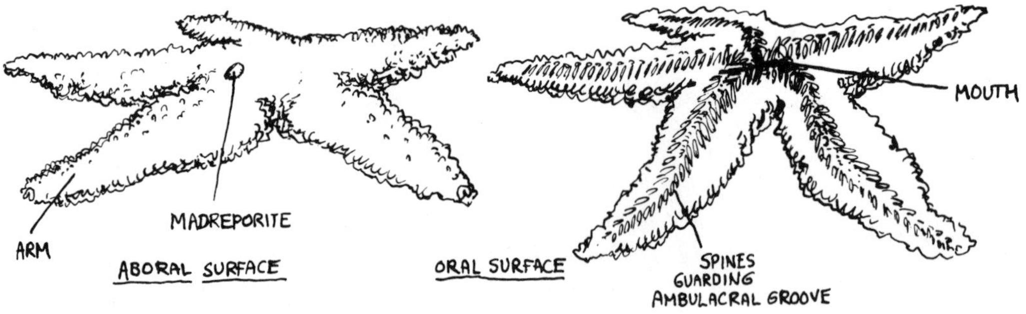

First, look at the knobby, spiny exterior, which gives the phylum its name, Echino (spiny) + derm (skin). In some forms, the spines can be quite long. Among other things, they make echinoderms very unpleasant to eat. Notice also the fivefold pattern, not only in the number of arms (which may be greater), but in the grooves and rows of spines and plates. This pattern is a unique echinoderm characteristic.

Look at the "bottom"—the *oral surface*—of the starfish for the deep *ambulacral grooves* running down the arms. In them, probably retracted, are rows of *tube feet*—hundreds of stalked suction cups by which the starfish clings to the rocks and its prey.

At the tip of each arm is a small tentacle (frequently lost in preserved specimens) and an eyespot. On the aboral surface is a low, buttonlike structure called the *madreporite.*

Turn the starfish aboral side up, and using small, strong scissors, cut off the upper surface of one arm and the body, avoiding the area of the madreporite. You are cutting through a skeleton made of thousands of tiny bones, each a crystal secreted by a single cell, jointed to its neighbors to form a structure like a geodesic dome. (Imagine a starfish with sore joints!) As usual, be careful not to cut too deeply or poke the scissors point inward too far. Before you remove the upper surface you have cut off, fill your dissecting pan with water and immerse the specimen. Now, starting from the end of the arm, peel the surface off gently, teasing it free from underlying attachments with closed scissors points. When you reach the central body, fold the arm flap back out of the way, but try to leave it connected.

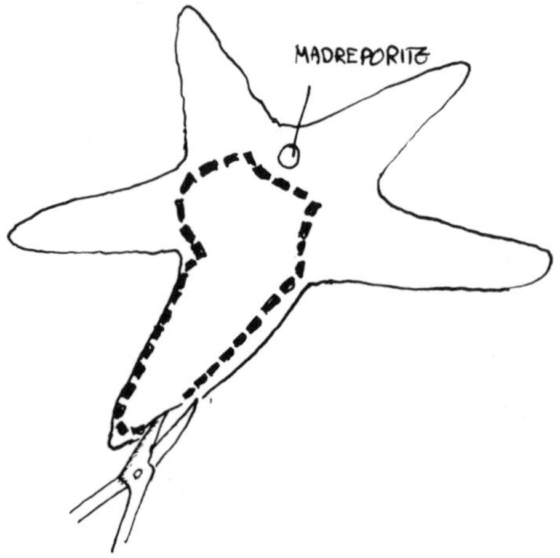

Most of the arm is filled with brownish *pyloric caeca,* which are glands that produce digestive enzymes and also absorb food once it is digested. Note the many fingerlike lobes, providing a great deal of surface area for all this activity. In the center of the mass you can see a thin, pale pair of main ducts; trace them back to the body, where they enter the *pyloric stomach,* a fairly flat, thin pouch. This is connected by a very short intestine (remember, absorption occurs in the caeca) to the anus, which is on the aboral surface. Currents created by cilia sweep the starfish clean. Below the pyloric stomach is the more complex *cardiac stomach,* very folded.

The cardiac stomach can be forced inside out and protruded from the mouth, and digestive juice from the caeca dribbles along its folds to reduce the starfish's prey to seafood chowder, which can be sucked up. Some starfish can slip the cardiac stomach between the shells of bivalves which have been wired shut; presumably the irregularities in the shells provide enough gap.

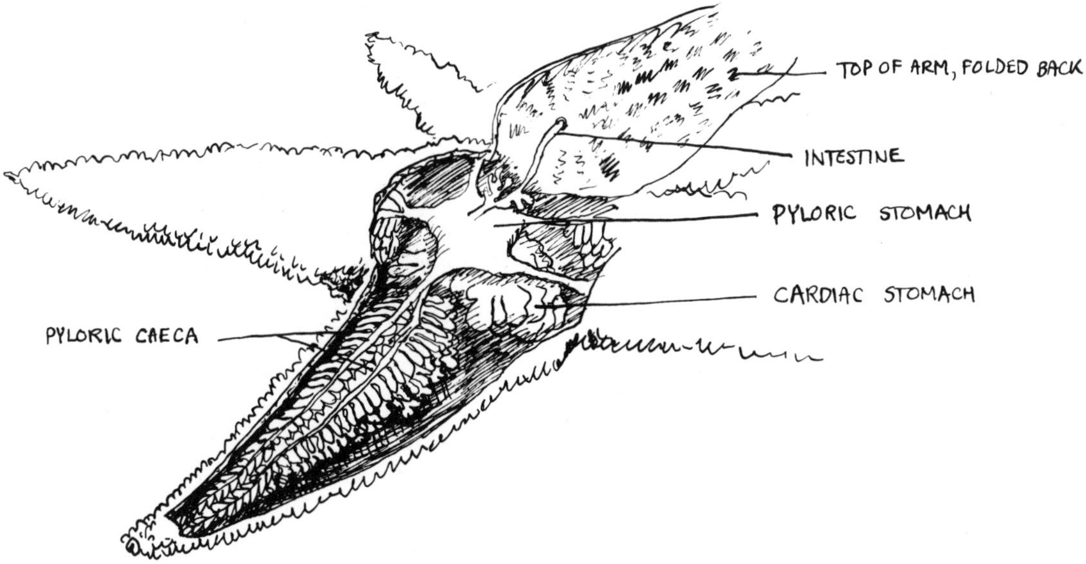

Remove the pyloric caeca from the arm (you will have to pick out many of the brownish bits, for the organs are likely to break up), and remove the pyloric and cardiac stomachs by cutting away the upper parts and carefully picking away the lower ones. You should see the structures shown below.

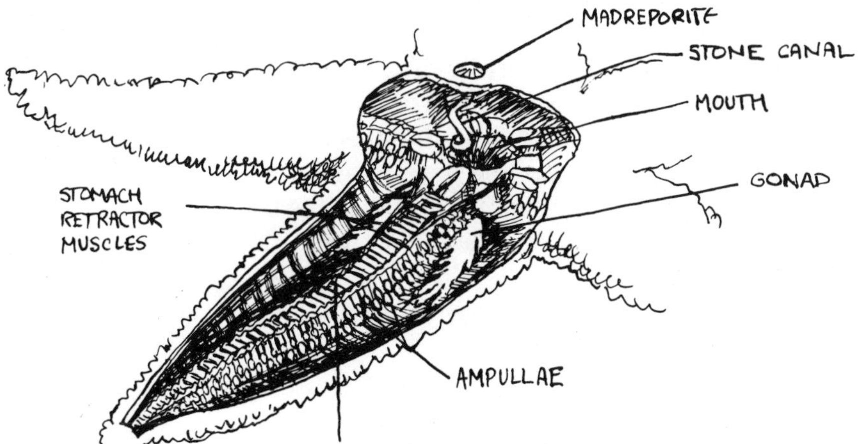

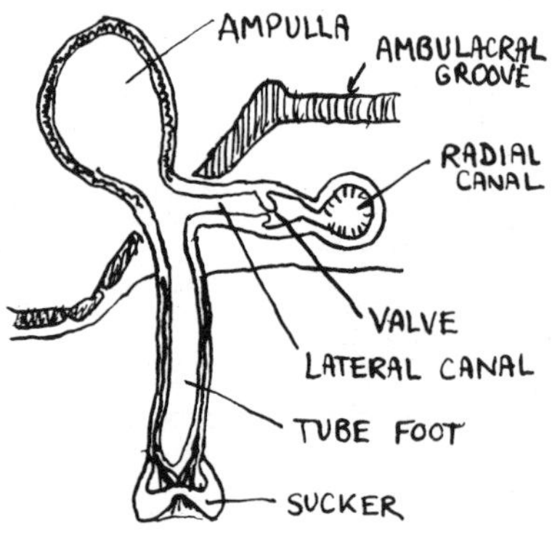

A SINGLE TUBE FOOT IN CROSS-SECTION

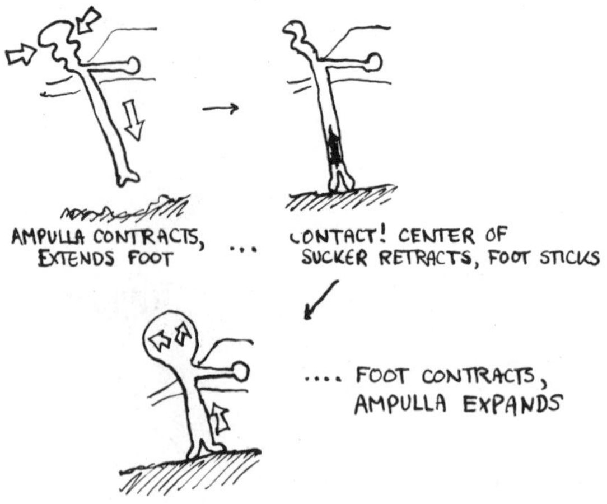

AMPULLA CONTRACTS, EXTENDS FOOT ...
CONTACT! CENTER OF SUCKER RETRACTS, FOOT STICKS
.... FOOT CONTRACTS, AMPULLA EXPANDS

In particular, you will see five prominent toothlike ossicles (the name for the starfish bones) in a ring around the mouth; extending from each of these is a corrugated-looking ridge of smaller ossicles, which is the roof of the ambulacral groove. Under the madreporite you can see a white, bony tube, the *stone canal,* running down to the mouth.

This sets the stage for an account of starfish locomotion, a process that involves a *water vascular system.* Use a hand lens or dissecting microscope to examine the surface of the madreporite; it is, as you see, a strainer. The stone canal contains sea water, which can pass in through the strainer and, via the stone canal, down to a circular canal surrounding the mouth (this is buried below the toothlike ossicles). The circular canal connects with a cilia-lined radial canal running just under the ambulacral ossicles in each arm. Each tube foot is connected by a short, lateral canal to the radial canal as shown above.

You can see the sacs, or *ampullae,* along the sides of the ambulacral ossicles. The tube feet emerge in the ambulacral groove. The rest is hidden. But just try gently squeezing the ampulla while watching the tube foot connected to it! (This is most easily done on a slice of arm just a few millimeters thick, under a lens.) The ampulla, in life, is contractile and forces the foot to extend (and makes it rigid); muscles in the foot work the sucker and shorten the foot.

Buried in tissue just around the mouth is a nerve ring, the closest starfish come to a brain; if it is cut, the tube feet on two sides of the animal may begin moving steadily—and powerfully—apart, with disastrous consequences. But then, given the starfish's regenerative power, it isn't as bad as all that!

Questions:

1. In what way is the starfish skeleton unlike that of other invertebrates?
2. Why do the pyloric caeca extend so far down each arm? (Hint: are you aware of any organ system you haven't seen in the starfish?)

The Producers

Being animals ourselves, we tend to find other animals interesting. Plants, on the other hand, usually elicit a sigh of boredom. They are, at best, a sort of background to our lives. Nothing could be more unfair! Plants are not only attractive, they are also essential to our life as the ultimate source of all of our food and most of our oxygen. They also present us with mysterious and exciting physiological problems, for they have systems very different from ours: no muscles, but they can move; no nerves, but they can react to stimuli. Their use of sexual reproduction is more complex than ours, and in fact gives rise to some of students' greatest problems in learning about plants. These strange, vital organisms are much more appealing than you think, so be prepared to look at your potted geranium with a new respect and wonder.

What, exactly, *is* a plant? That's not too easy to answer, once you get beyond things like redwoods and tulips. In fact, the whole group may not be such a useful one after all, and in the five-kingdom scheme used here, there are three kingdoms—Monera (prokaryotes), Protista, and Fungi—that used to be partly or entirely considered plants. We will look at some representatives of each kingdom, attempting to travel from simple forms to more complex ones, and later give more detailed considerations to "true" plants.

Prokaryotic Plants

Besides the bacteria, the kingdom Monera contains the *cyanobacteria*, or blue-green algae. These are single cells (or groups of single cells) with a basic structure like that of bacteria. But their cell membrane is greatly folded, and in the folds lie pigments like chlorophyll and carotenes; with these pigments the blue-greens trap sunlight and use its energy to build up a form of starch, so these are producers, "feeding" only on water, carbon dioxide, and solar energy. They are the type of life that has been around the longest—fossils several *billion* years old strongly resemble modern blue-green algae. These old-timers are still very much with us, for every summer they grow in fresh-water ponds and reservoirs, emitting a characteristic stench that makes drinking water revolting (although

not toxic). Easily identified by their deep blue-green to black color when seen *en masse*, cyanobacteria also have some distinctive features when seen under the microscope. Often they are joined in long strands or large globs, glued together with a gelatinous substance.

Make a wet mount of *Oscillatoria*, a filamentous blue-green. Examine it under high power. Who said that plants don't move! This strange, corkscrew motion, like that of many blue-green algae, has not yet been adequately explained—there are no cilia or flagella involved! In some of these filamentous blue-greens, there is a certain amount of specialization of cells—some being nutritive, and some reproductive.

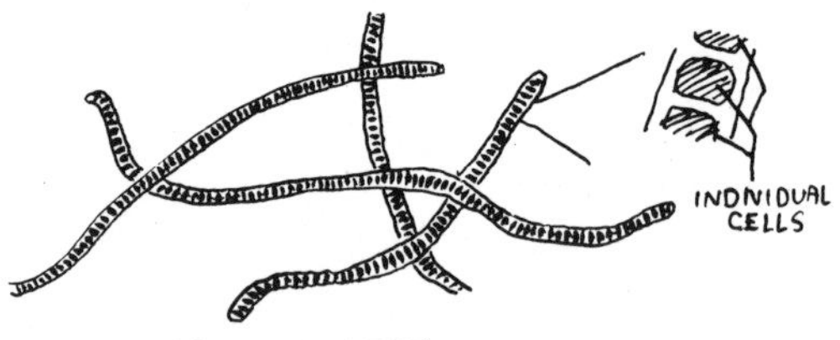

OSCILLATORIA

Protist Plants

Other producers have internal membranes but are free-living single cells—protists. You have encountered *Euglena* and the diatoms already. These are photosynthetic cells with true chloroplasts containing various light-trapping pigments of many colors. Take a look now at a third protist type, a dinoflagellate. This creature has a rather unusual nucleus, one with only a single nuclear membrane and chromosomes that are permanently condensed. Externally, you can see two flagella—one extended, and one wrapped around the organism like a belt, lying in a deep groove. The dinoflagellates move about in the water (like *Euglena*) to get the best light conditions; their numerous chloroplasts function best at certain light intensities. The role of these protists in the ecology of the oceans is a major one, both because they are an important food source and because they can sometimes produce a nerve poison that can kill huge numbers of fish in a "red tide."

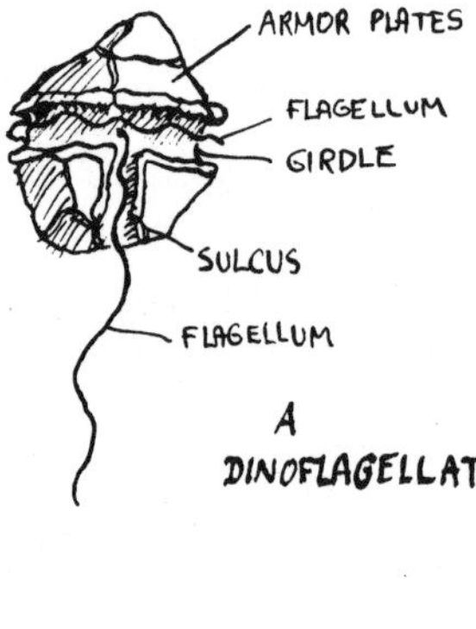

A DINOFLAGELLATE

Fungi—The Dropouts

Every scheme of organization has its weak spot, and that spot is frequently the fungi. These were considered degenerate plants for a long time, perhaps algae that had lost their chloroplasts. But they resemble flagellates in some ways, and their cell walls are chemically similar to arthropods' exoskeletons! Naturally, in the five-kingdom system they are simply given their own kingdom. And they certainly don't belong in this exercise, "The Producers," for they *aren't* producers, but feed on decaying plants and animals or parasitize living ones. Still, they are plantlike in many ways, and they will help to introduce us to a major plant characteristic, the alternation of generations.

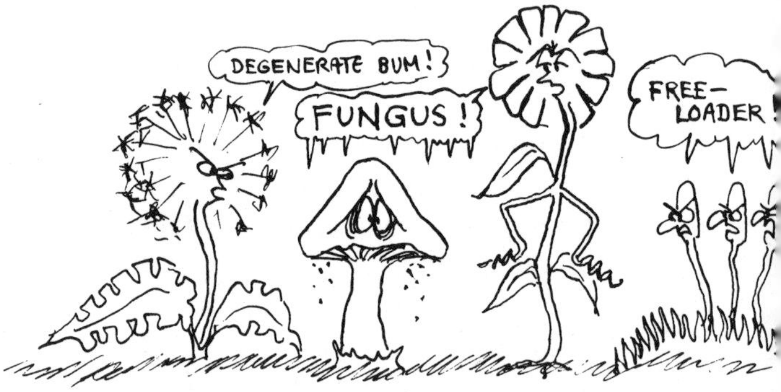

So, with those reservations in mind, prepare to consider the fungi. We will start with single cells—the yeasts.

Yeasts are single-celled organisms, mostly harmless decomposers (though some kinds can cause infections). They reproduce by a kind of mitosis in which one daughter cell gets less cytoplasm and so must do more growing; this is called *budding*. If this were the only way yeast reproduced, life would be simpler. But, in reality, yeast with the usual number of chromosomes (diploid) can divide by meiosis to produce yeast with half that number (haploid), and the resulting cells go on dividing mitotically for any length of time! That is, there are really two kinds of yeast, haploid and diploid. And the haploid yeast comes in two "mating types," which can—you guessed it!—fuse to produce a diploid yeast again. It's as if our eggs and sperm

set up on their own and propagated more of themselves before fertilization. Why all this? Well, in an uncertain world, it allows a better chance for sexual reproduction, for one thing.

If you put a pinch of bakers' yeast into sugar water, it will soon expand and bubble, as anyone who has made bread knows. It is fermenting the sugar (six carbons, you remember, though in table sugar this is linked to a five-carbon sugar, which is broken off first) and releasing CO_2 (the bubbles) and—if and when the available oxygen is used up—ethanol (two carbons), the alcohol found in wine and beer.

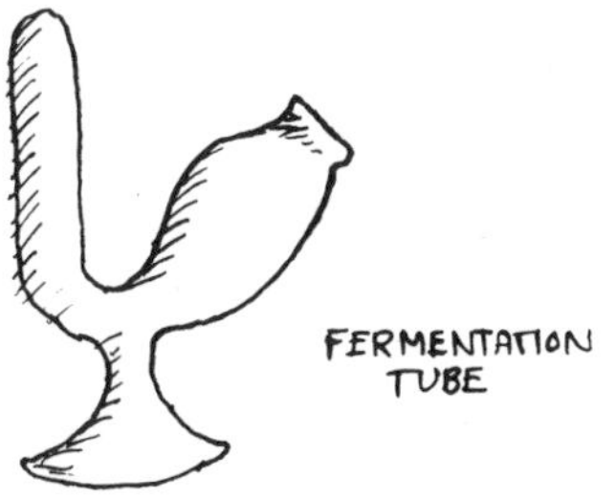

If you can get a fermentation tube—a piece of glassware with a long finger to trap gas, such as the one shown here—you can compare the ability of yeast to ferment different sugars. Simply make a solution of the desired sugar, about 5% in strength (that is, 5 g in 100 mL water), and add several drops of a solution of an indicator dye such as phenol red. If the solution is yellow, add 0.1 *N* NaOH drop by drop until it turns red; if it is purplish, add 0.1 *N* HCl drop by drop until it turns red. Then fill the fermentation tubes as shown by your instructor, add a pinch of yeast to each, place them in a warm spot, and leave them while you complete this exercise. For sugars, try sucrose, lactose (milk sugar), maltose, and others—you might be curious enough to try sugar substitutes like saccharin. At the end of the period, note in which tubes gas has accumulated and acid has been formed (turning the red dye yellow), signs that yeast is using the sugar. Also sample some growing yeast from your tubes, adding a drop of methylene blue stain to a drop of yeast culture on a microscope slide. Under high power you can see budding forms and the distinctive, oval shape of yeast cells.

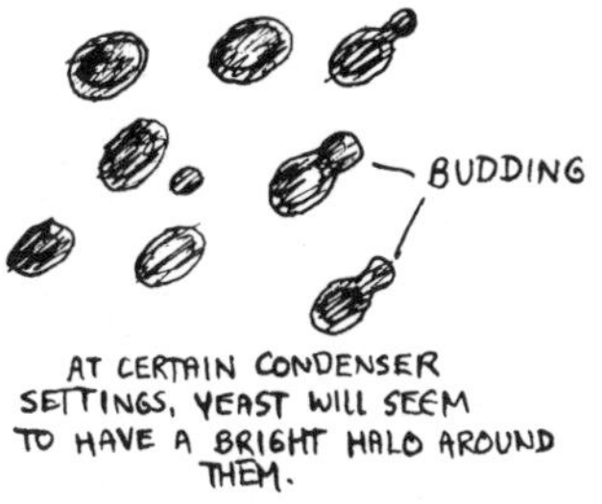

BUDDING
AT CERTAIN CONDENSER SETTINGS, YEAST WILL SEEM TO HAVE A BRIGHT HALO AROUND THEM.

Fungi come in various forms, and many are multicellular: there can be quite a bit of specialization, especially in the reproductive structures. Use a needle to pick up a bit of bread mold or another kind of mold (try to get some of the darkest part). Mount this in water (no stain is needed) on a microscope slide, add a coverslip, and examine on low power, with your condenser somewhat stopped down. The mass of filaments you see are *hyphae* (high′-fee), each one a large *syncytium* (many nuclei in one cell: sin-sish′-um). These act like roots, diving into the bread and releasing enzymes to digest it, then absorbing the nutrients. On stalks above the hyphae you will find dark, spherical *sporangia* (if you have a mold other than the common bread mold, these may have various shapes and colors). These contain asexually produced spores, which can be spread widely when the sporangium pops. Crush one to see the spores. Now, here's the shocker: everything you are seeing is *haploid*. Once in a while, two hyphae of opposite

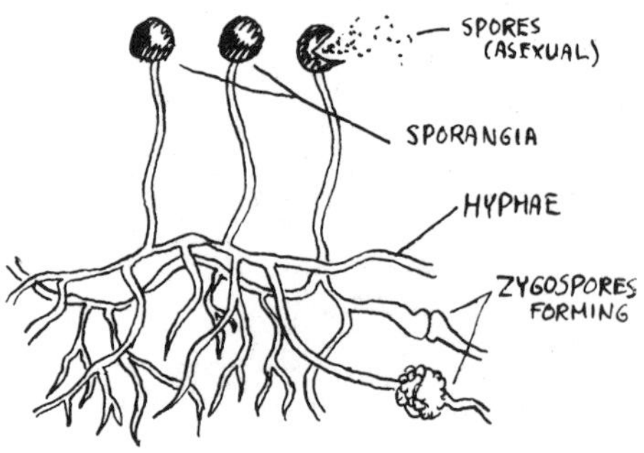

mating types meet, fuse, and form a spherical, diploid *zygospore;* but when this grows, it undergoes meiosis so that only haploid hyphae emerge from it! The diploid cells (like *us*) have become a fleeting stage in an essentially haploid organism (like our sex cells)!

True Plants—The Green Algae

The *green algae* are similar in some ways to the euglenids; many of them are flagellated, some have stigmata, and, of course, they are green. But they usually have only one huge, cup-shaped chloroplast, and, most strikingly, they usually have cell walls. These rigid envelopes allow the cells to dispense with contractile vacuoles—they can only swell so much!—but they cannot flex and curl like euglenids. There is always a price for wearing a suit of armor. But for sheer beauty as well as a thought-provoking lifestyle, consider the green alga *Volvox.* If you can get any *Chlamydamonas,* look at these cells first—you will see rounded versions of the euglenids, with rigid walls and two flagella. Now look at *Volvox.* Aha! The *Chlamydamonas*-like cells have produced a clear jelly sphere—a hollow one—in which they have embedded themselves. Thin processes reach from cell to cell in the manner of struts—a geodesic sphere. And, in beautiful coordination (how do you suppose they do it?), each cell's twin flagella beat together with all the others, making the whole colony roll majestically along like some green space station. Young daughter colonies grow up in the sphere's interior and eventually burst from this glass womb to go solemnly tumbling off on their own. If you watch long enough, you're liable to be hypnotized. For sheer, lovely oddness, nothing beats *Volvox.* Yet you, multicelled creature that you are, should feel a twinge of familiarity looking at all those cells working together. On such hangs our larger style of life.

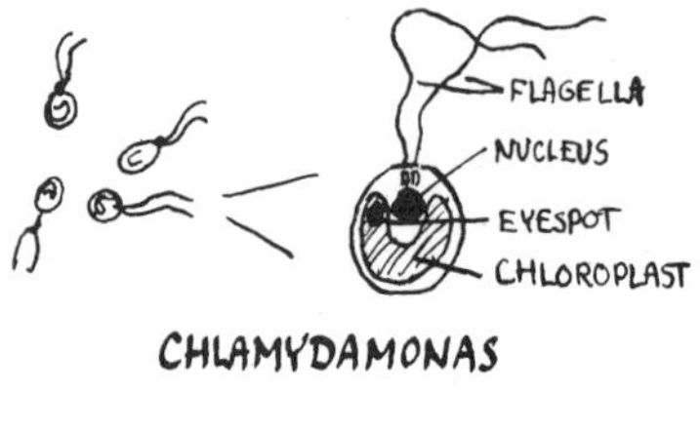

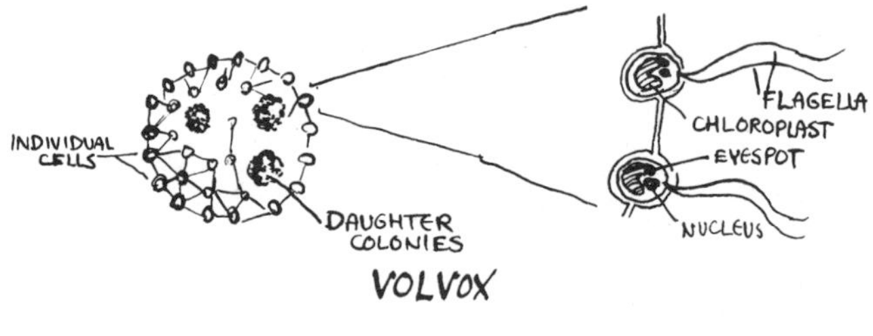

The Terranauts

The plants we have considered up until now have been aquatic (except for the nonplants, the fungi—and even these do require a damp environment). Now, the ocean has its advantages, such as a fairly stable temperature and strong currents to ride in. But for a plant, gas (CO_2) and light are essential, and these are in short supply anywhere but in certain restricted layers of the ocean. And even at best, the available light is not that great, as any underwater photographer knows. But on land—ah! Paradise!—blinding sunlight and CO_2-laden breezes. When plants began to expand their territory from water to land, there was only one catch: there was precious little water out there. Plants have to have some water for photosynthesis, but that essential amount is relatively small. The history of plants' conquest of the land is a history of the reduction of their water requirement to that bare minimum: the war against drying. Little battles in that war are still being won, for it remains one of the major limits on plants today. Maybe that rustling in the leaves you hear is really a round of applause for the latest triumph.

Lichens

We can't bring up the topic of water requirements without taking a glance at a bizarre solution to it—a very old sort of symbiosis, the composite organism called a lichen (and often miscalled a "moss"). On any walk through the woods you will see them growing on stones (and slowly crumbling them)—flat, pale-colored blotches with wavy edges. It is a double organism: a fungus and an alga, in a permanent and mutually beneficial relationship. You may have sections of lichens to examine; identify the parts shown in the following drawing.

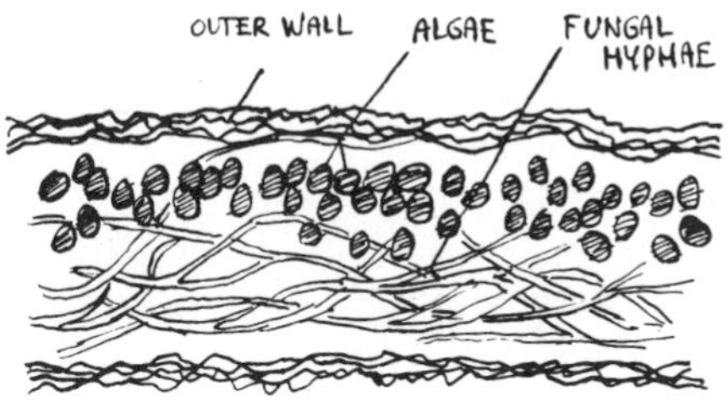

This represents a solution to the drying problem; the tough-walled fungus encloses and protects the algae; in their artificially moist environment, the algae do what the fungus can't—trap solar energy and synthesize nutrients, which are then available to both organisms. This is one way of "solving" the drying problem: getting someone else to solve it for you!

Mosses

The first group of plants to become truly adapted to land—the first "terranauts"—were the Bryophytes, a group including several obscure types like liverworts and a more common division, the mosses. A liverwort looks rather like a flat scrap of stranded seaweed. Its ability to live in areas that are moist but not submerged is due to the development of a specialized layer of cells, an *epidermis,* which slows down evaporation. So, once again, an environmental challenge too great for single cells or colonies of similar cells is met by specialization. We have different tissues, now, for protective as well as reproductive purposes.

Mosses have begun to deal with one other problem facing the terranauts—support. Seaweed can be as long as a building and still wave gently in the water—aided, at most, by gas-filled floats. On land, the same tall plant would flop down flat. And that would mean that its light-trapping surface—and therefore its energy source—would be no greater than the ground area it covered. Oh, to have more surface area!

The mosses have managed that; they have stalks with a central stiff cylinder of supportive cells; arranged around the stalk are flat *leaf scales,* which are the sites of photosynthesis,

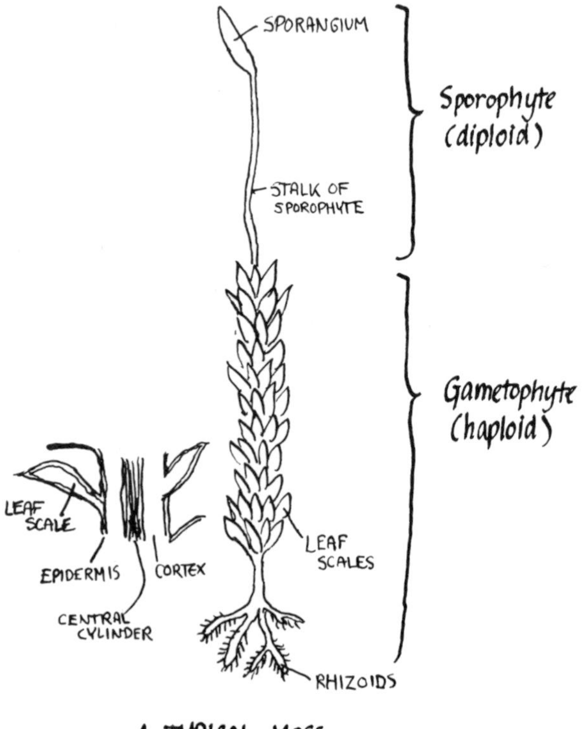

A TYPICAL MOSS

and below are strands of cells forming *rhizoids,* which anchor the moss and absorb water and nutrients. Use a dissecting microscope to identify the parts shown.

Note that the leafy part of the moss is haploid (as were fungal hyphae). In the plant kingdom, this is called the *gametophyte* (compare with our haploid "gametes"). This is either male or female, and at its tip produces correspondingly, an *antheridium* or an *archegonium.* The antheridium produces flagellated sperm, which, if the weather is wet enough, swim in a film of water and climb into the nearest archegonium, where they fuse with an egg and form a diploid organism, the *sporophyte*—which grows right out of the archegonium, forms a long stalk, and develops a spore sac, or *sporangium,* at its tip. From this come (of course) *spores.* Open one and see.

The gametophyte moss is not usually very resistant to drying, but it can come back after drying out. The sporophyte has a thick, waxy cuticle (feel the difference in hardness and slickness), which makes photosynthesis difficult but prevents drying. So the sporophyte lives as a parasite on the gametophyte and is able to produce tough spores. There are two major limitations to the system: the sperm require a film of water to reach the egg, and, for reasons we will see, mosses can get only so big.

The Advantages of Plumbing

If you remember the animal phyla, it probably doesn't shock you that mosses can only get so big. Diffusion imposes limits on size—how could enough water get to the leaves or enough sugar to the roots of a large plant, if it had to get there by the same process that spreads the sugar around your unstirred tea?

Animals, you recall, installed plumbing. Well, so did plants! All the rest of the plants we will consider are *vascular plants,* with elaborate systems of tubes connecting all parts. As you might expect, however, the forces that move fluids in these tubes are not at all like the ones that act in animals.

We will begin with a simple demonstration of these vascular systems. Prepare a strong solution of a dye in water—food coloring, ink, or a biological dye such as methylene blue, fuchsin, or malachite green. The dye should be so concentrated that it looks black. Now procure a stalk of celery, or, to be fancy, a white flower like a daisy or a carnation. Use a sharp, clean razor blade to slice across its stalk, about 8 inches from the top, and put it into a flask or vase full of the dye. Seal around the stem with paraffin film or plastic wrap, and then put the plant upright, in a sunny window or under lights, and in a breeze. Check on what is happening to the plant over the course of this exercise, and if necessary, let the demonstration go on overnight.

Vascular plants have the typical organs—roots, stems, leaves—all riddled with vascular tissue and all forming a closed system, coated with a waxy cuticle. This, you will remember, made problems for the moss's sporophyte; photosynthesis requires gas exchange, but what is watertight is generally airtight, too. The solution may be seen in this way:

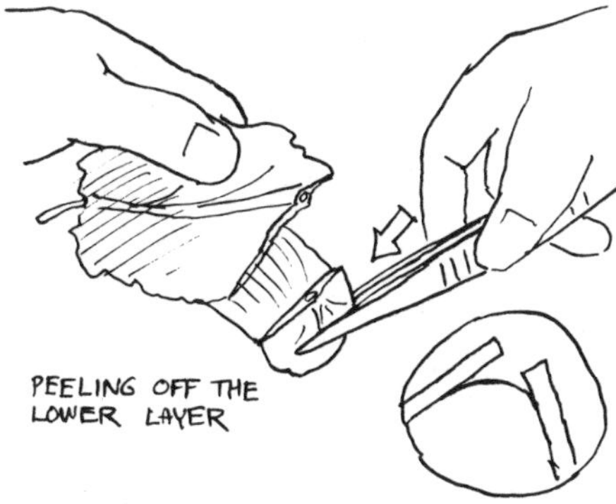

Nick the upper surface of a leaf (geranium or other thick leaf is best) and snap and pull it as shown at right (rather like stringing beans). If you did it right, the lower epidermis should come away as a clear sheet. You only need a bit of it. Put this in a drop of water on a slide, flat; add a coverslip and tap out any air bubbles. Now examine under low power, with the condenser

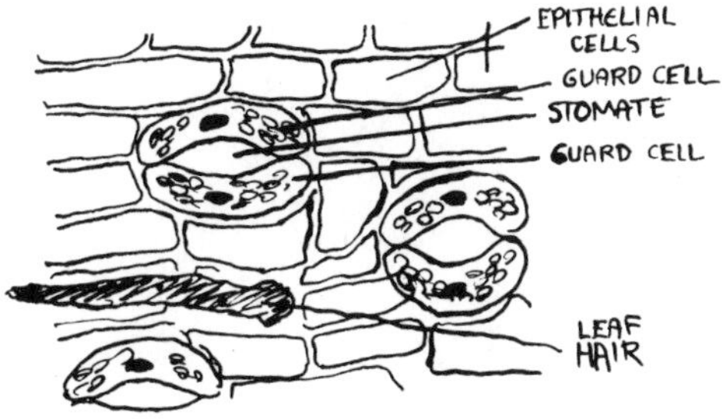

slightly stopped down. Most of the epidermal cells are visible only as the boxlike shape of the cell wall, but you will also see stomata and guard cells. The guard cells can actively open and close the stomata—enough for gas exchange without excessive drying!

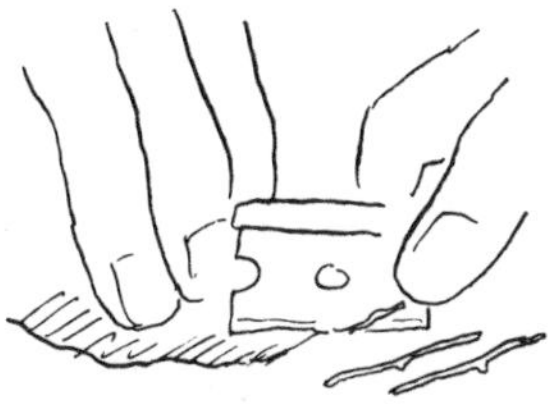

You can use the same leaf to make a section; just steady the leaf with one hand and shave off the thinnest slice you can, making a number of slices and selecting the best. Flip this slice, cut side down, into a drop of water on a slide and examine it under low and medium power. Identify the layers drawn below.

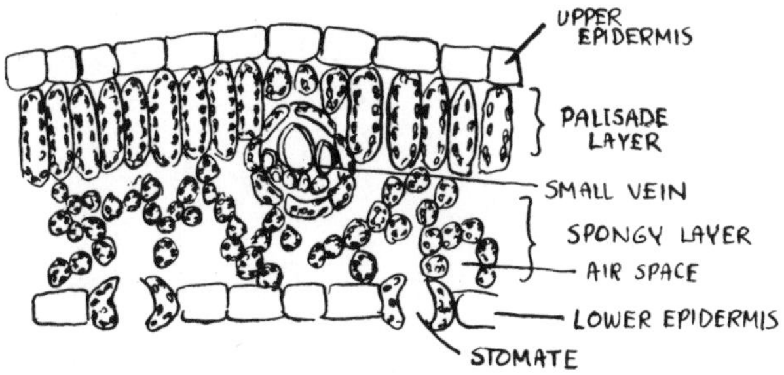

Here's a great light-trapping device: a clear, waxy upper epidermis, then close-packed palisade cells arranged to soak up the light from one end to the other; a spongy layer to let gas exchange freely, and the lower epidermis with its controlled openings. And also a vein, which brings us to vascular tissue again. This is best seen in a section of a stem. Prepared slides will show this best, for instance a stained section of a buttercup stem. At this point, some of the differences among vascular plants show up; we will confine ourselves to the flowering plants, which will show one of two basic arrangements of vascular tissue.

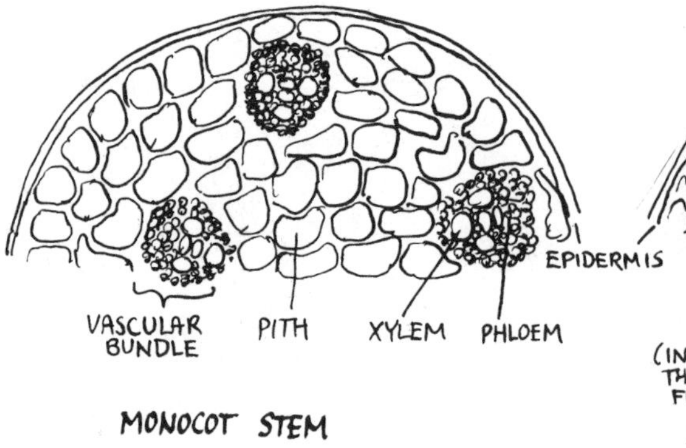

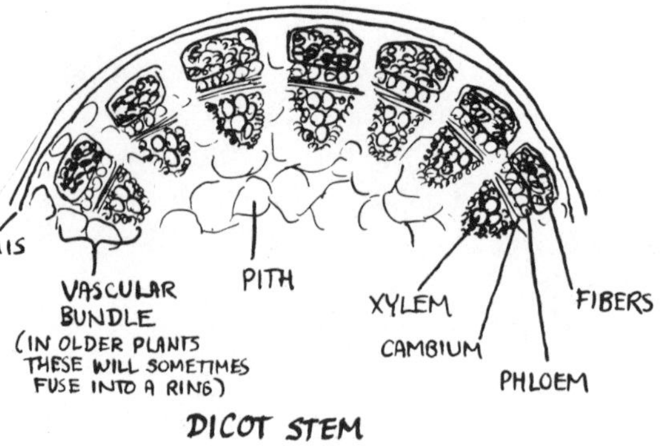

Notice that most of the stem is supporting tissue—pith and the like. The vascular tissue lies in dense clumps and is mainly composed of *xylem* (zye′-lum) and *phloem* (flo′-um)—the former, large tubes formed from the empty walls of dead cells, and responsible for conducting water to the leaves; the latter, smaller tubes with sievelike partitions, still living tissue, and responsible for distributing sugar formed in the leaves to all parts of the plant.

How do fluids move in these tubes? Fast and forcefully—too much so to be satisfactorily explained by any theory as yet. But it seems clear that evaporation from the leaves, and plant cells' ability to manipulate osmotic pressure, have something to do with it. Have you got a blue celery/daisy by now? *Did* evaporation affect the process? As a last step, slice a thin section of the stem above where it was immersed and verify the identify of the type of tissue that transported the colored water.

The Ubiquitous Ones: Prokaryotes

The first person to see the bacteria that surround us was a nearsighted Dutch cloth merchant, Antonj van Leeuwenhoek. He made microscopes as a hobby, very simple ones with a single, tiny, but powerful lens. One day, he scraped his teeth and found that his mouth had a greater population than the city of Delft. Like most curious people he was not revolted by this (though he protested, in letters, that he brushed his teeth every day), and he described what he saw very well. He didn't know it, but he was getting the first look at a world of organisms that cause diseases, nourish us, and live with us most intimately.

It was a while before he was believed, partly because his observations were hard to duplicate. Besides, who wanted to think of having a mouthful of living creatures? It was a hundred years before anyone seriously studied these creatures, and another hundred before people became generally aware of them. They were found to be everywhere, and due to the work of Koch and Pasteur, some of them were found to cause diseases. So when they finally became generally known, "germs" were cast as the heavies. People went around wearing surgical masks and even refusing to shake hands. We are just getting over this now. Microbes are good, bad, and indifferent; some of the ones in our colon could kill us if they got anywhere else, but would also make us very sick if they disappeared from the colon. Most of the ones you will see are harmless, although if you have any cuts you shouldn't get any of these microbes in them. What we want to illustrate here is the variety and habitat of bacteria.

Bacterial cells are fairly simple in structure; unlike other cells, they have no internal membranes. Their DNA is naked (no histones) and simply sits in the middle of the cell, on the *mesosome*, an infolding of the cell membrane. They are full of ribosomes (slightly smaller than eukaryotic ones) and may have a crystal of bacte-

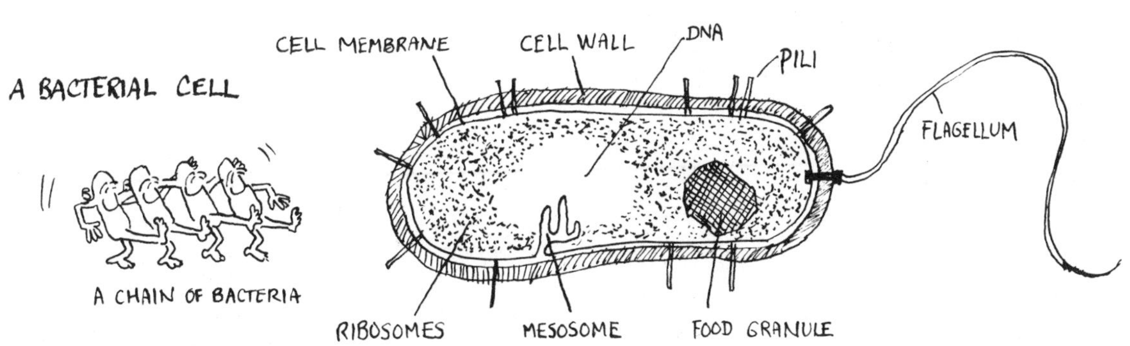

rial starch or minerals in storage. Their membrane is coated by a cell wall consisting of sugars and amino acids forming a sort of close-woven fabric which can be unraveled by certain antibiotics, leaving the bacteria exposed and vulnerable.

Bacteria may propel themselves by beating flagella, but these are not quite like protists' flagella—bacteria have a rotating joint (the only one in nature) at the flagellar base, and *spin* their flagella. Pili are short, sticky appendages that help anchor bacteria to surfaces and to each other.

To begin with, take a clean toothpick and scrape the surface of your back teeth, especially between the teeth and near the gums. Smear the whitish material out in a small drop of methylene blue on a microscope slide, cover with a coverslip, and look under low power for the bluest areas; then switch to high power and observe. If you had a very large drop, you will have to wait until everything stops sloshing about. Then, the crucial question: is your mouth as populous as van Leeuwenhoek's?

In looking for bacteria, you must remember to look for very *tiny* but very *regular* forms, like those below:

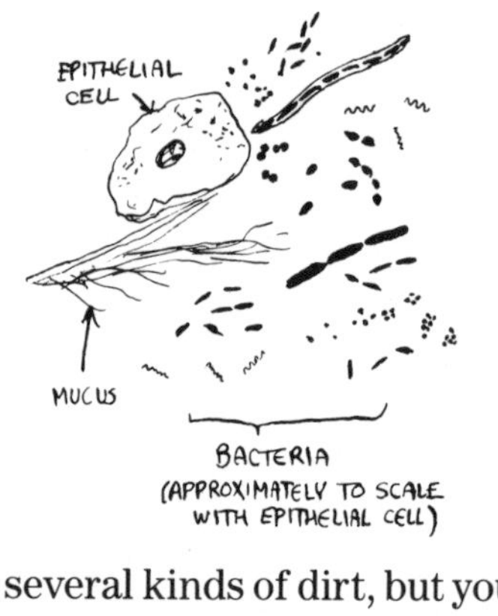

You may see several kinds of dirt, but you will soon notice that the bacteria have a special look to them. They will all stain quite deeply with this method, and they are almost always present in large numbers of the same kind. You can begin to classify bacteria by their shapes: bacilli (buh-sill′-eye), cocci (cock′-sigh), and spirilla (spuh-ril′-la). These are shown below:

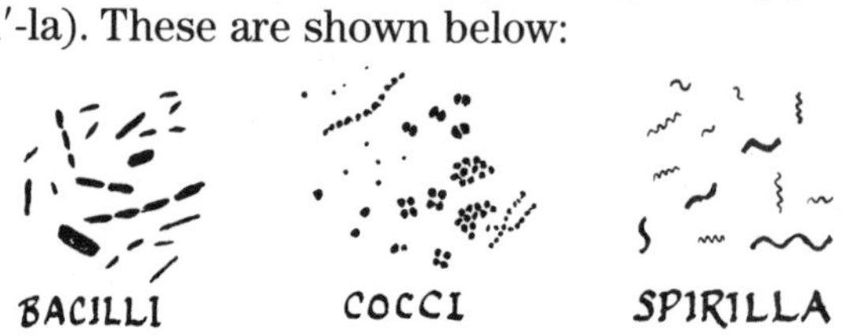

Bacilli are rods, longer than they are wide. They may be very tiny or very large (relatively!); they may form chains, the daughter cells remaining joined together. Cocci are spherical and often form chains (streptococci) or clusters (staphylococci) or sometimes just pairs (diplococci). Spirilla are wavy, and the rarest kind, although they are often present in small numbers in the mouth; they are often very slender. There are a few other shapes, such as ovals (coccobacilli) and commas (vibrios), but you're not likely to see any. And just to make life interesting—and textbooks outdated—a *square* bacterium has been reported.

To make a more permanent slide, as well as to see the bacteria more clearly, you can do a simple stain or a Gram stain. The Gram stain takes only a few minutes longer and provides some crucial information about the bacteria. When a Danish microbiologist spilled alcohol on his stained smears, he noticed that some bacteria bleached out quite easily with this solvent, while others did not. It turns out that this is due to a fundamental difference in the cell wall structure of the two types of bacteria and is the single most important characteristic after shape for identifying most bacteria. (In addition, many antibiotics only work on one kind or the other.)

To do either stain, first make a smear; tranfer a drop or so of water to a very clean microscope slide. The drop should be no wider than a dime and quite thin; if you are using bacterial loops, two loopfuls is plenty. Then take a tiny speck of the whitish material from your teeth, or a tiny speck of bacterial colony from a plate, and swirl it into the drop until you have a faintly milky, even suspension. Allow this to air dry thoroughly. Then pass the slide through a burner flame as shown in the illustration, moving it slowly but keeping it in motion. If you stop moving and hold the slide in the flame, it will crack.

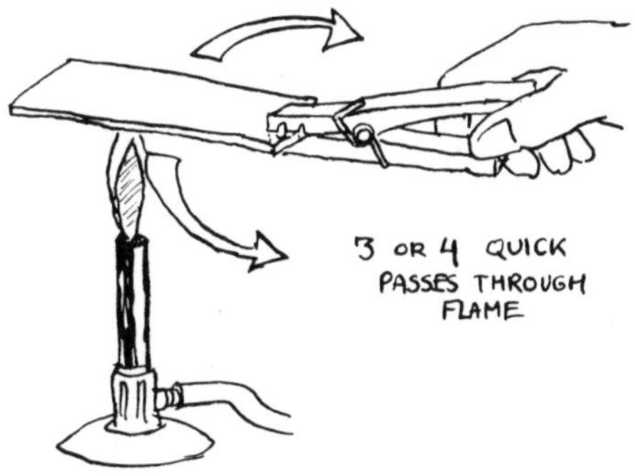

To do the simple stain, get any bacterial stain (methylene blue is a common one) and, when the slide has cooled, form a pool of stain over the dried smear. Let it sit for a minute or two and then carefully wash off the stain with running water. Let the slide air dry; it is possible to blot it very carefully with absorbent paper, but don't move the slide or the paper sideways at all or you will scrape off a lot of bacteria. Examine the slide under oil immersion.

To do the Gram stain, let your smear cool and then cover the smear with a pool of crystal violet. After one minute, rinse off the stain and replace it with Gram's iodine. Allow this to sit one minute, and then rinse off the iodine. Tilt the slide and gently trickle acetone-alcohol over it. This step should be done quickly, taking no more than 10 seconds or so. You will see some stain bleed off most slides at this point, and as soon as the alcohol runs clear rather than stained, it is finished. Rinse the slide immediately with water. Then apply safranin stain and allow it to stand one minute, rinse the slide off, and air dry it. Examine it under oil immersion.

With the Gram stain, bacteria that are Gram-positive will look dark purple, while Gram-negative bacteria will look rose to pink. You may see some ambiguous straining; many bacilli seem Gram-negative when they are old but Gram-positive when young, and some cells stain in a spotty way with some areas of each in every cell.

Now we will try to prove the exercise's title. Bacteria can be sampled from the environment in various ways, and *counted* . . . without a microscope. Impossible, you say? You have a hard enough time *with* a microscope? But Herr Doktor Robert Koch (Kok, unless you speak German and can make a throat-clearing sound at the end) is—or was—one step ahead. He realized that one bacterium, if it fell on a nutritious, solid surface, would divide and divide until a mob of its descendants made a visible glob—a *colony.* (They can't walk away.) You have a plate of such a medium with 150 colonies on it? You must have started with 150 bacterial founders, one for each. (These, by the way, are also *clones.*)

All the great man needed was a jellylike, solid medium. But all he had was gelatin, and anybody who eats in a cafeteria knows how solid *that* is, when warm. Then the wife of one of his students, a widely traveled woman, recalled a seaweed product that makes a firm jelly even when warm: agar. This, cooked up with beef broth, made a nutritious, solid medium for bacterial growth which has never been improved on. There are many variations on it; you will use one called *nutrient agar.*

The agar comes in flat glass plates called *petri dishes* (now often made of plastic). You may need to pour in the melted agar yourself—if so, fill the plates only halfway, and keep your hands away from it. Or the plates may be already poured. But in either case, try to open them *only* as shown below, using the lid as a shield. Avoid leaning, or passing your hands, directly over the agar surface. Try, also, to keep the plates out of drafts when they are open.

Your task, with these plates and a few supplies such as sterile swabs, beakers, and sterile water, will be sampling your environment for bacteria. As time and supplies permit, do one or more of the following. Be sure to keep careful records of what you did, and to label the plates clearly.

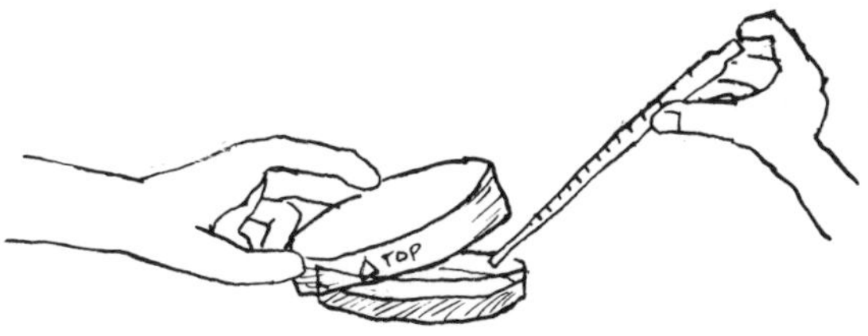

Do one or more of these (on separate plates):

1. Swab a surface area as follows:
   a) Use a sterile swab moistened with sterile water.
   b) Firmly rub a measured area (say, 10 cm by 10 cm) with it, working all over as if scrubbing it, for at least 1 minute.
   c) Try doorknobs and faucet handles, especially in rest rooms; food preparation and serving areas, especially salad bars (try swabbing a lettuce leaf!); pillowcases and sheets, and so forth. Just keep a record of exactly what you did.
   d) Put the swab in a sterile container (test tube) if you have to carry it around; keep it moist. Wipe it back and forth gently on the agar surface in the petri plate.
2. Cough, sneeze, or talk into an open plate 15 cm away.
3. Wash your hands (without soap) in 300 mL of sterile water. Use a pipette to transfer 1 mL of this to a plate, and rock the plate around to distribute it evenly. (Remember that what will grow is 1/300 of what was on your hands.)
4. Allow an insect to walk around on the agar surface for a set length of time—say, 10 minutes. Flies and cockroaches are good for this, ants not so good. Tarantulas are great!
5. Lean over the plate and run your hands through your hair.
6. Swab your mouth, gums, or throat gently and treat sample as in 1*(d)*.

Others will occur to you; one student the author knows got amazing results from her lipstick. Be creative!

Results

After incubating the plates for a day or two (at room temperature, with the plates stacked upside down to prevent condensation from dripping on the agar), examine them. Note particularly the size and appearance of the different colonies, and record the number of each kind. Sometimes, if the plate was too wet, you will see an overall growth; this can be described but not counted. If time permits, use a toothpick to sample a colony, and make a smear and stain it as above.

Questions:

1. Calculate, for any plates in (1) or (3), above, the total number of bacteria present on the surface sampled (this may involve estimating total surface areas of complex shapes).

2. What is the "dirtiest" thing sampled? The "cleanest"? Why, do you suppose?

The Tools of the Trade

There was once a time when, if you had a university degree, you could explain to anyone who would listen just exactly what the whole world was all about. The nature of light, the inner workings of the brain, and the fate of the universe—any of these could be your topic. Your ability to push your theories was limited only by the strength of your voice. You could be a sort of oracle or guru.

But then some students not unlike yourselves began to resist this authority-centered approach. They insisted on putting theories to the test. Any theory, they said, which is worth listening to should make some predictions about some observable things—and, if these things are not observed, the theory simply must go! This shocking idea was too much for some people who felt that it reduced a scientist from being a genteel philosopher to being a "mere mechanic." But ever since then, people in science have been sending millions of beautiful theories to their doom.

In fact, the very term *theory* is reserved, strictly speaking, for complex systems of explanation which have withstood the test, and *hypothesis* or lesser terms are used to name these poor expendables. At a scientific meeting the air is swarming with them; they are introduced with a mere "What if . . . ?" rather than a name. But if it is your very own shiny new hypothesis, launched into the void, it is hard to see it shot down, and still harder to do it yourself.

The hardest experiment to do is the one that will bring your pet hypothesis down like a house of cards, and it is a common observation that researchers take days and days getting ready to do this kind of experiment—in truth, working themselves up to it. There is also a tendency to squint hard at the results, trying to make them say what you want them to. If you want to avoid this problem, of course, you can always make hypotheses that lead only to vague predictions so that nothing can disprove them! (A lot of pseudo-scientific cults use this technique.)

To help get around this tendency to squint, scientists insist on quantitation. Bluntly put, if you can't measure something, you can't claim to know anything about it! The poet in us immediately rebels. How can you measure love? Or truth? Or even the pain of a toothache? But fortunately science isn't—for most scientists—a philosophy of life, but only a set of rules for finding out how things work. And even the most ardently subjective poet wants an exact pound of cheese when it has been paid for, and tends to

look askance at publishers who make accounting errors! We all agree that *some* objectivity is possible and desirable.

So, basic to all the sciences, we find measuring devices of all sorts of levels of complexity, from rulers to mass spectrometers. They are all ways of quantitating the world. Biologists have the hardest time with quantitation because living things have a habit of walking away—or worse—when you're trying to quantitate them!

Seriously, the extreme variability of living things forces biologists to be particularly aware of the limitations and proper use of all means of quantitation. This exercise is designed to expose you to various tools of measurement; some are familiar, some not. Try in each case to get a feel for the purpose and limitations of the method.

A Note on the Metric System

In America, the metric system still has a forbidding, mysterious aura. This is perpetuated by those idiotic brochures that inform you that there are 2.113436 pints in a liter. (When was the last time you measured a millionth of a pint, anyway?) For practical purposes, you need something like this:

A *meter* is about one yard (and a kilometer is about 2/3 mile)
A *liter* is about one quart
A *kilogram* is about two pounds

These aren't conversion factors, but rather "reasonableness" factors to give you a general idea of the size of the units. If you want accurate metric measurements, you simply use a metric measuring device; conversion factors are useful only in converting old-style recipes and such into metric units, something to be done only once and written down.

Once you are in the metric way of thinking, all you need to know is the system of prefixes for larger or smaller units, of which only the following are commonly used:

kilo- = 1,000; deci- = 1/10; milli- = 1/1,000; micro- = 1/1,000,000.

(In talking about very small units sometimes nano-, for one billionth, is also used.) These few ideas are all you need to get started, and if you keep them in mind you won't sit down in a restaurant and order a kiloliter of wine, or go into a fabric store and order a decimeter of fabric for a dress.

Linear Measure

The most common measurement is a linear one, and we all think we are experts at this. But people like field biologists, who must estimate size and distance routinely, say that most people are poor at this. In fact, it is commonly observed that reports of animals sighted by nonnaturalists consistently overestimate the animal's size by a factor of two. Can you do better?

Your instructor will show you several animal specimens of types unfamiliar to you, for only 30 seconds or so. These will be in different places in the room or, if possible, out of doors. You will try to estimate two things: the *size* of the specimen and its *distance* from you. You will probably give your estimates in the more familiar inches and feet, but if time permits, your instructor may ask you to try metric units as well. Compare your estimates with reality by making measurements with a meter stick. What was your most common kind of error? Practice until you can see some improvement. Note that this is really much easier than such estimates on live animals; for one thing, movement can make a difference. It is claimed that a bear moving toward the observer appears much larger than one moving away. To go from the large to the small, we turn to the microscope as a measuring tool. You are not likely to have this on your microscope, but you may see a demonstration of an *ocular micrometer* (mike-rohm′-et-er, to distinguish it from 1/1000 of a millimeter). This is just a slip of glass with a scale etched on it, often numbered from 1 to 5. Five *what*, you ask? Aha!

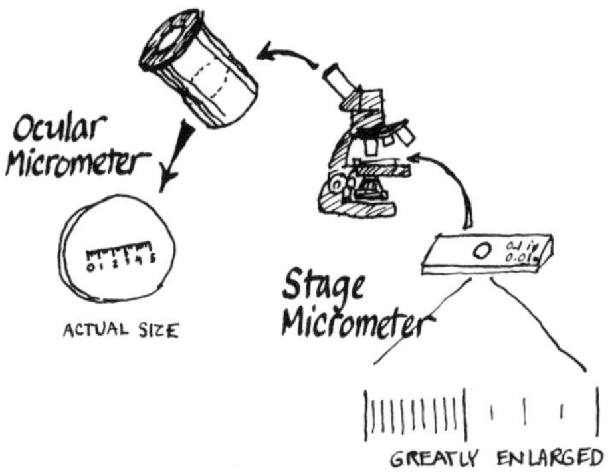
Ocular Micrometer
ACTUAL SIZE
Stage Micrometer
GREATLY ENLARGED

This is an important concept, so your innocent question has triggered a lecture. You see, although some people call them "micrometer units" to give them a name, these 5 divisions by themselves *have no meaning.* They do not correspond to any particular measurement like millimeters or inches. They are just reference marks. What "5" means in terms of length will vary with the magnification of the object. Any such scale (and they are very common) must be *calibrated* to give measurements in meaningful units like millimeters. This is done by comparing a known standard to the scale under the same conditions under which you wish to make measurements; in this case, at the same magnification. You can then assign values to the units on your scale and measure anything with it—provided the conditions don't change. The way this works out in the case of the ocular micrometer is this: you obtain a *stage micrometer* that has lines on it 0.1 millimeters (mm) apart, and some lines 0.01 mm apart. This stage micrometer is a thick slide on which the microscope can be focused. In focus at some particular magnification, you move the slide so that the scale of the ocular micrometer is seen superimposed on the scale of the stage micrometer. You then say, "Aha! 4.7 ocular micrometer units just span 0.1 mm, so any future reading in these units, multiplied by 0.1/4.7, will give me a length in mm."

If possible, go through this calibration process for the demonstration microscope and then check your accuracy by measuring the diameter of red blood cells in a stained smear of human blood. These cells should turn out to be 0.0074 mm, or 7.4 micrometers, in diameter. How close are your measurements? After a little of this you might want to know why anyone in his or her right mind would *want* to measure a red cell, anyway. Among other things, certain anemias, particularly megaloblastic anemia, are distinguished by a slight increase in the size of the red cells; here, a small measurement can make a big difference.

Weight and Volume

We make rough estimates of weight and volume all the time. Our measuring devices are inaccurate and our methods are sloppy, but it usually doesn't matter. Too much salt in the soup? Just throw in extra potatoes. But too much salt in the intravenous drip in a hospitalized patient can have consequences that are much more difficult to undo. For biological purposes as well, measurements of weight and volume must be good. But what do we mean by "good"? Chiefly, we mean two things: *accurate*, conforming to the real, universal standard; and *precise*, able to be repeated and to give the same results consistently.

HELLO, FRED— TELL THE PHARMACY THEY JUST PICKLED ANOTHER ONE!

Balances, as the name implies, allow you to determine weight by a direct comparison of your sample to known weights. *Scales* generally use a spring that is stretched or compressed by the weight of the sample, the spring having been calibrated with known weights in the factory. Scales are cheap to produce and easy to use, and so are common for home use; but as anyone who has ever slept on an old mattress knows, springs change their response to weight over time. Laboratories use balances in preference to scales.

The heart of a balance is a knife-edged pivot (one or, usually, more) on which the beam moves. It is cruel and unusual punishment to allow a balance to bang back and forth, thus dulling these pivots. Some balances have a lever that can be thrown to support the beam; with others, you must support the heavier pan by hand until balance is nearly reached. Many balances have sliding weights for grams and multiple of grams, and separate weights for larger values or for tare weights (weights to counterbalance a container, to save the trouble of subtracting the container's weight from the total). Balances for weighing very small amounts (such as milligrams) have glass enclosures and a mechanism for loading weights so that you never touch them; at this level, moisture from skin can make a significant difference.

Volume is measured simply in calibrated containers of various kinds: *volumetric flasks* having a single marking to denote one particular volume, *graduated cylinders* with many markings like measuring cups, but tall and narrow, and, for small volumes (10 mL or less), *pipettes* —thin, graduated tubes. In glass containers, water forms a concave surface, a *meniscus;* and all glass measuring devices are designed to be read from the bottom of this, as shown above, right.

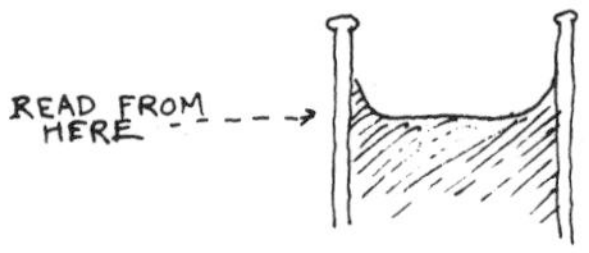

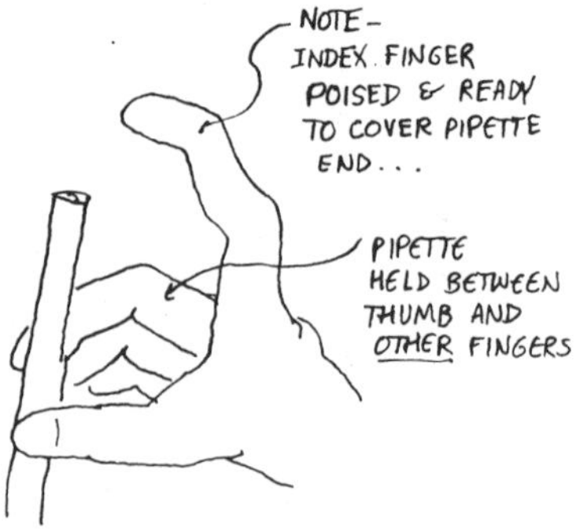

Using a pipette involves a few unusual techniques. First, pick up the pipette by its wide end and hold it as shown. Then place the narrow end below the surface of your solution, put the wide end in your mouth, and suck up the solution until it is above the graduated part of the pipette. Now comes the hard part: get your index finger firmly placed over the wide end of the pipette just as you remove it from your mouth. If you are too slow, the solution will slip down into the graduated region. Practice until you can stop it in time. To release the solution from the pipette rapidly, just raise your index finger; to release it slowly, slightly ease the pressure on the index finger while slowly rotating the pipette, using a back-and-forth movement of your thumb and other fingers. Do this until you can stop and start the flow at will. For toxic or otherwise dangerous substances you will substitute a rubber bulb or other suction device for your mouth in this process.

To measure out a volume of liquid with a pipette, first select the right size pipette. Most mistakes in pipetting come from the user picking up a 0.1-mL pipette instead of a 10.0-mL one, even in very advanced work. There are many sizes, often kept together, and although some pipettes are color-coded by size, not all are. You must look at the pipette above the graduations, where you will find the total capacity in mL and the size of the smallest divisions in mL, written as "1.0 in 0.01" and so forth. Pick a pipette whose total capacity is close to the volume you want to measure. Fill it as above, and then slowly let out solution until the bottom of the meniscus is at

0—at the *top* of the graduations, you notice; the scale is upside down compared to other measuring devices. Now place the tip over the container you want to put liquid into, and slowly release solution until the bottom of the meniscus is at the exact volume you want. Remember how much the smallest divisions represent! As for the solution left in the pipette, either return it to its source or discard it. If you want the whole volume contained in the pipette, most biological pipettes allow you to empty the pipette by gravity, holding the pipette vertical and touching the side of the container; their tops are marked by "TD." If the remaining material is to be blown

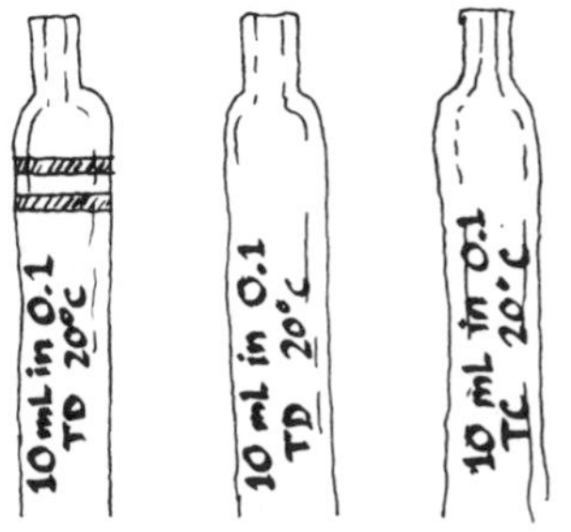

out, such TD pipettes are marked with one or two bands, as shown at the far left of the illustration; if not, no bands are present, as at center. A third kind of pipette, less common than these, is marked "TC" and is designed to be not only blown out but also rinsed to catch every last bit. This is most often used with sticky fluids like blood.

Probably you will encounter only the pipettes with "TD" and two bands, but you ought to know that others exist.

For a practice exercise with pipetting and weighing, use the old standby: weighing water. One of the glories of the metric system is that units are even and symmetrical, and one milliliter of water weighs one gram. For small volumes, weight can be more accurately determined than volume, and it is common for the accuracy of automatic pipetting devices to be checked by weighing water pipetted by them. (You would be shocked at the size of the errors made by some very advanced devices!) Of course, evaporation will take a toll, so weigh quickly and keep out of drafts. And the 1 g/mL figure is for water at 4° C, so there is some temperature effect, but at room temperature (18–20° C) water is only about one percent lighter per milliliter. Still, pipette some water for your partner to weigh, and then trade places. Repeat one volume several times: what is the average weight per milliliter? How does this compare to that given above? By how much did the repeated readings differ?

Color and Light

A good wine taster can tell a lot about a wine by its color, and most of us can at least tell a red wine from a white! When you hold a glass of burgundy up to the light, white light enters it but red light comes through it. The other colors (all there, of course, in white light) were filtered out on the way through; there was something in the wine that absorbed them. A rosé wine has some of the same stuff, but not as much as burgundy, so it appears less red. You apply this rule—that the more of a colored substance there is, the more light it absorbs—every time you look at a cup of coffee or tea and declare it too strong or too weak. In the laboratory, the process is made quantitative with a *spectrophotometer.* Color is specified exactly as a particular wavelength of light, in nanometers; this wavelength is separated from the others as in a rainbow, and the amount of this light which passes through the sample is quantitated by a photomultiplier, which is similar to a light meter on a camera. Voila! We have numbers! But it will never catch on with wine tasters.*

*One of my students has, since this was written, gone into the field of enology, or wine research, and she informs me that her research does, indeed, use the spectrophotometer to measure the amount of pigment in wine.

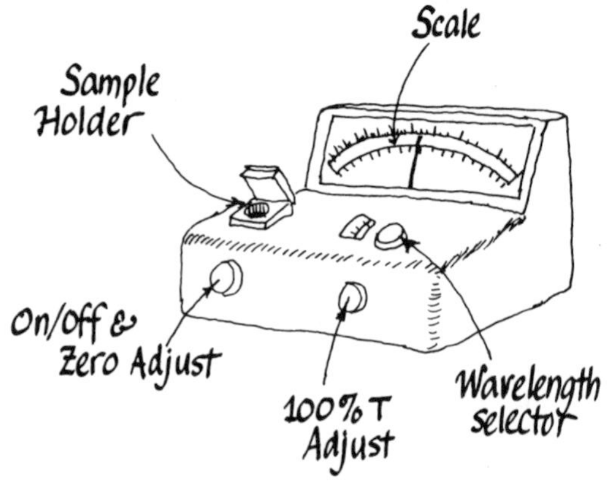

A TYPICAL STUDENT SPECTROPHOTOMETER

You will be given a solution of a colored compound in water, plain water, two test tubes (or possibly special containers called *cuvettes*), and access to an instrument probably resembling the one shown. You will need to know the wavelength at which your compound absorbs light best; set the wavelength selector to this number. Turn the spectrophotometer on and let it sit for 10–15 minutes unless it has already been warmed up (newer models may not need this). Then adjust the zero adjustment until the needle on the scale is at 0% T on the left-hand end of the scale. This sets total darkness at 0. Then fill one tube with plain water and, after wiping it clean and dry, open the lid of the sample holder, slide the tube in, and close the lid. Now use the 100% T adjustment to place the needle at 100% T at the right-hand end of the scale. You have just told the instrument to consider this water perfectly clear—no matter what may really be in it. You may now put your solution into the other tube and replace the water tube with this new sample tube. Now what does the scale read? This indicates how much light your sample absorbed.

Dilute your sample, using your best pipetting technique. You might try these dilutions: sample, 1 mL, water, 2 mL; sample, 1.5 mL, water, 1.5 mL, sample, 2.0 mL, water, 1.0 mL, and so forth. The precise sequence doesn't matter as long as you keep a record of what you did. Then record the absorbance and transmittance of these—that is, record the number from the % T scale and the one from the A or OD scale, usually printed right above the % T scale or right below it. Be careful in reading this scale. Do this for 5 different dilutions.

Now use graph paper to plot your data. Call the concentration of your original solution 1.0; your dilutions are then

$$\frac{\text{Volume of Sample}}{\text{Vol. Sample} + \text{Vol. Water}}.$$

Plot these values on the X-axis.

Plot your % T readings on the Y axis, and, on a duplicate graph, your A values on the Y axis. Which one gives a straight line? Can you derive a simple relationship of concentration to A or % T? Could you determine the concentration of an unknown dilution from this? If time permits, exchange unknown dilutions with other groups (*you* keep a record of the dilution) and see whether you can do this. This is in fact one of the most powerful and widely used tools for the determination of unknown quantities of a large number of substances, both clinically and in research.

Questions:

1. What range of concentrations of your sample can the instrument detect? What could you do with samples outside the range?
2. Get the actual concentration of the sample from your instructor and rework your "simple relationship" in terms of moles. Why is this more valuable?

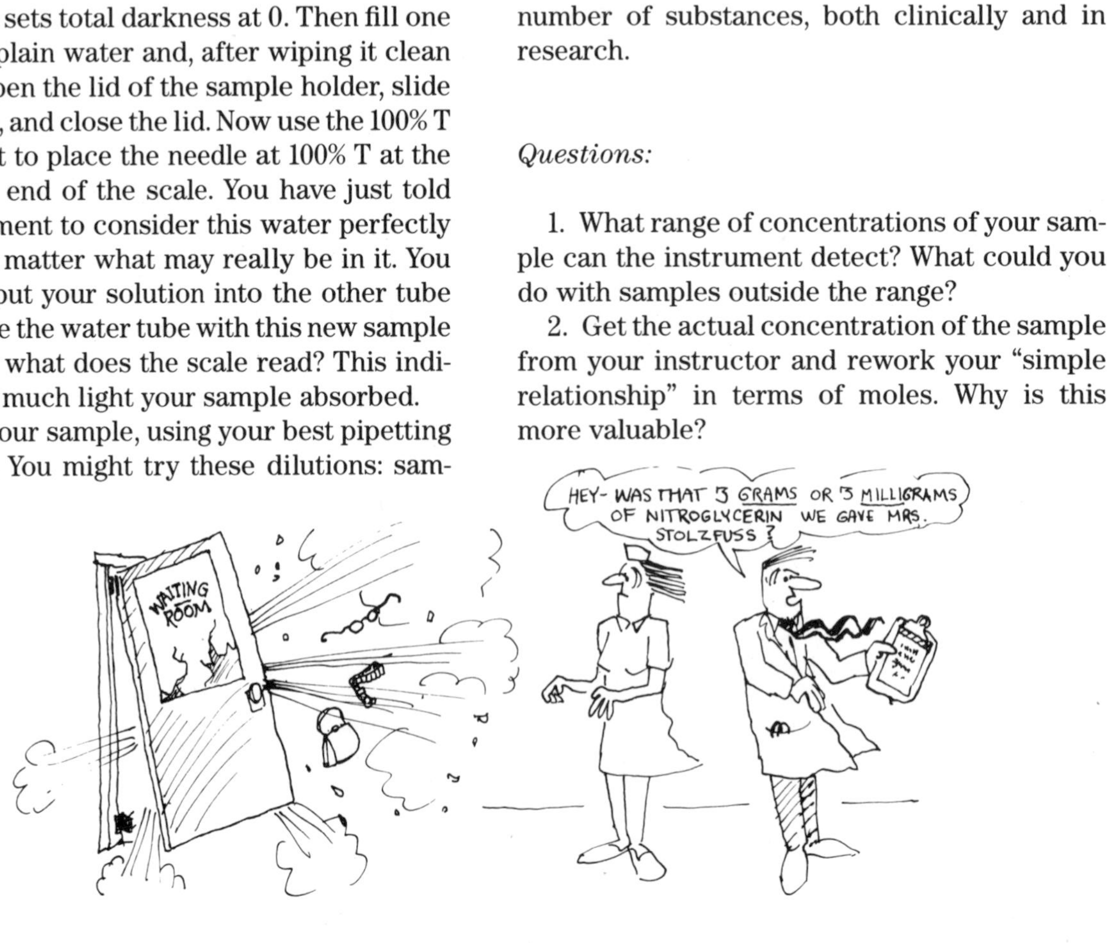

The Cell's Alchemists

Chemistry, you know, has very disreputable origins. The first chemists were obsessed with transforming one substance into another, usually a less valuable substance into a more valuable one. Because often the things they were trying to transform were elements, they didn't get too far. But along the way they made a lot of useful observations; some things *could* be transformed. Heat cinnabar (red mercuric oxide), and presto!—droplets of metallic mercury appear. Many such reactions were catalogued, and eventually people started noticing that some substances could speed up a reaction without actually being changed or used up themselves. Magic! These useful things were called *catalysts.* It seems that they work by providing a perfect place for the reacting molecules to get together—a chemist's catalyst is rather like a matchmaker's intimate, candlelit restaurant.

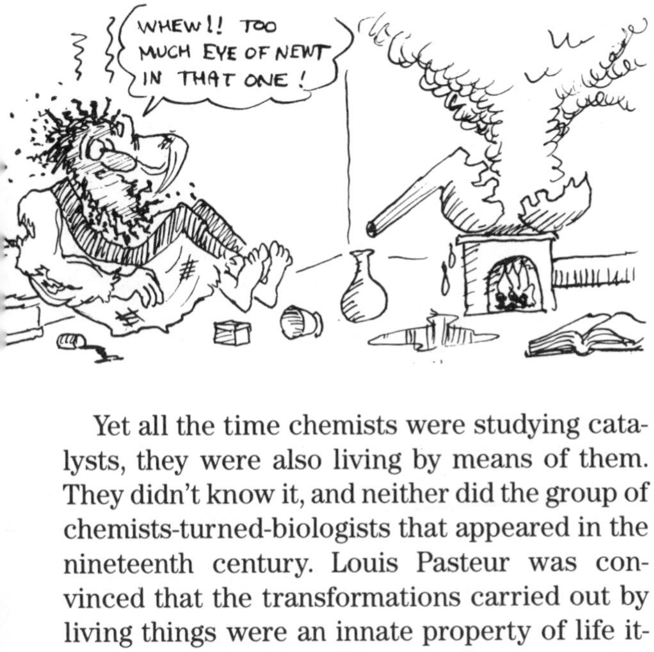

Yet all the time chemists were studying catalysts, they were also living by means of them. They didn't know it, and neither did the group of chemists-turned-biologists that appeared in the nineteenth century. Louis Pasteur was convinced that the transformations carried out by living things were an innate property of life itself, inseparable from life. Only live yeast cells,

for example, could carry out the intricate process of fermentation, that critical part of winemaking which Pasteur had studied so thoroughly. The name for all such transforming principles, in fact, was *enzyme*—literally, "in yeast." The shock came when Edward Buchner made an extract of yeast with the same transforming activity.

Enzymes, apparently, were just catalysts (though very good catalysts) and could be taken right out of cells.

For some people, the magic was gone. For others, it was just beginning. The great enterprise, which is *very* far from finished, is the effort to account for life processes in chemical terms. Every year, the charts showing the chemical reactions carried out by living cells become more and more complex, and one bit of the puzzle fits into place. Enzymes catalyze almost all of these reactions, from the trivial to the vital. (Several kinds of severe mental retardation, for instance, are the result of a single defective enzyme.) We will look at the activity of one enzyme, whose activity is easily observed, since it catalyzes the formation of a colored product.

The Basic Reaction

Our enzyme is found in many plants and is responsible for the darkening of the cut surfaces of many fruits and vegetables. Originally, enzymes were given rather picturesque names, like Old Yellow Enzyme; but now the names give the organic chemical name of the substrate or product (substrate is the starting material) and an "-ase"ending to show that it is an enzyme. Sometimes the terms *oxidase* or *reductase* are used when the reaction is an oxidation or a reduction.

Our enzyme is polyphenol oxidase, and it converts the −OH groups on catechol into =O groups, producing benzoquinone, which is colored reddish brown. To do this, the enzyme requires oxygen; and, along with benzoquinone, it releases water.

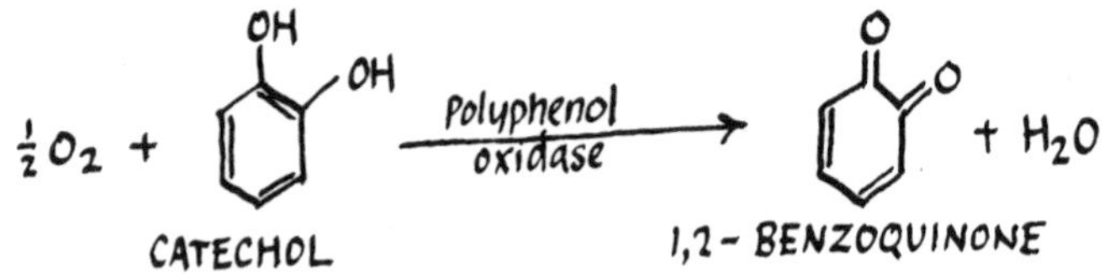

Now, why would a plant do anything like this? It is a response to injury. There is a little catechol around in most plant tissues, and when the tissues are cut open and exposed to air, the enzyme catalyzes the reaction shown and produces benzoquinone, which is a natural antiseptic. (It is too irritating to use on people, however.) So the browning of cut fruit is a protection against disease.

Your instructor will make a crude potato extract, a potato in water, pureed and strained. Keep it on ice when it is not in use. This is your enzyme solution. There will also be a solution of catechol, the substrate. Keep this on ice, too. Try not to cross-contaminate solutions (by switching from one to the other with the same pipette).

Set up three small test tubes, labeled 1, 2, and 3, as follows:

Tube 1
1 mL enzyme solution
1 mL water

Tube 2
1 mL water
1 mL substrate solution

Tube 3
1 mL enzyme solution
1 mL substrate solution

Put them in a water bath set at 37° C for 10 minutes, giving them an occasional shake. Now why, you ask, did you do tubes 1 and 2? They are *controls*, without which no experiment has any value. Think: how do you know the crude enzyme solution won't turn red all by itself? How do you know the catechol solution won't? (In fact, they both will, if given long enough, though if everything is fresh and clean, you shouldn't see anything in ten minutes.) Mother Nature can fool you pretty thoroughly—and you had better watch out for your instructor, too. So you do *controls. Always.* Some textbooks talk about a "controlled experiment" as if there were other kinds—there are not, except in jokes and rejected papers. How do you select controls? Just use your imagination and try to think of every possible way things could go wrong; anything is a legitimate suspect.

By now, you have your first results. Describe the color of each tube. Does the enzyme work as advertised? Save tube 3.

Changing Conditions

Now we will try to learn some properties of enzymes by doing variations on this basic reaction scheme.

Enzymes are affected by the amount of substrate present; to determine how, set up tubes with different concentrations of catechol:

Tube 4
1 mL 1/10 substrate solution
1 mL enzyme solution

Tube 5
1 mL substrate solution
1 mL enzyme solution

Tube 6
1 mL 5 × substrate solution
1 mL enzyme solution

Tube 7
1 mL 10 × substrate solution
1 mL enzyme solution

Put in the different substrate solutions first, then add enzyme rapidly to all and put in the water bath. Watch the tubes, and note how long it takes each one to get as dark as tube 3.

Enzymes are temperature-sensitive; set up tubes 8, 9, 10, and 11 just like tube 3 or 5 above (all alike) and immediately put one in cold water in a refrigerator (10° C), one out on your bench (about 25° C), one in the water bath (37° C), and one in boiling water (100° C). This time, let all tubes go for 10 minutes, then rate them all on color—none, 0; little, +; more, + +; most, + + +. Explain your results.

Another factor in enzyme activity is pH. We could use different buffer solutions, but for simplicity's sake we will just add a small amount of acid or base. Set up the following tubes:

Tube 12
1 mL acetic acid solution
—
1 mL enzyme solution
1 mL substrate solution

Tube 13
0.1 mL acetic acid solution
0.9 mL water
1 mL enzyme solution
1 mL substrate solution

Tube 14
0.1 mL sodium hydroxide solution
0.9 mL water
1 mL enzyme solution
1 mL substrate solution

Tube 15
1 mL sodium hydroxide solution
1 mL enzyme solution
1 mL substrate solution

Tube 16
What should this tube contain?

If you said that tube 16 should be a control containing 1 mL water, 1 mL enzyme solution, and 1 mL substrate solution, go to the head of the class!

Now incubate these tubes for 10 minutes at 37° C. When the time is up, inspect the tubes and record their color, using the scoring system from the previous set of tubes. What can you say about the effect of pH on the activity of this enzyme? If time allows, use pH paper to estimate the pH in each tube, and plot color (0, +, + +, + + +) vs. pH.

Finally, enzymes usually require other molecules for activity—ions or small molecules, often vitamins—and these *cofactors* remain loosely attached to the enzyme. If something that can remove the cofactor is present, the enzyme cannot function. For example, EDTA (ethylenediaminetetraacetate) is frequently used as a food preservative because it can tightly bind calcium or magnesium ions; these ions are required by many bacterial and fungal enzymes. PTU (phenylthiourea) binds copper. As you have seen, one way to learn how an enzyme works is to learn how to *stop* it from working; so you should have no trouble understanding the purpose of these tubes:

Tube 17
1 mL EDTA solution
1 mL enzyme solution
1 mL substrate solution

Tube 18
1 mL PTV solution
1 mL enzyme solution
1 mL substrate solution

Tube 19
1 mL water
1 mL enzyme solution
1 mL substrate solution

Mix the first two ingredients and let the tubes stand 10 minutes, shaking them occasionally. Then add the substrate and put the tubes in the 37° C water bath for 10 minutes. After that, remove them and score them as before. What does this enzyme require as a cofactor?

So, born in alchemy, the study of chemical catalysts has shed a lot of light on life processes and has produced the field of enzymology. Enzymatic processes are so vital and so seemingly magical—perhaps all the alchemy isn't gone yet!

Questions:

1. Summarize the properties of this enzyme as you have observed them.
2. To keep fruit and vegetables from browning, people use lemon juice, bicarbonate, or citric acid. How do you think these work?

The Master Molecule

At the heart of modern biology lurks a most unusual molecule—DNA. You have probably heard about it so often that you take it for granted. But it is so unlikely a molecule that it was almost ignored.

DNA was discovered over a century ago by Friedrich Miescher, a young student in the world's first biochemistry laboratory. He wanted to know what was inside a cell's nucleus so badly that he collected pus-soaked bandages from a surgical clinic, to get lots of white blood cells. He extracted a new substance from the cells' nuclei; he called it *nuclein.* Some of it was protein, but some was a very peculiar acid, a polymer of four different subunits called *nucleotides.*

Now everyone knew that the stuff of heredity was to be found in the nucleus, but surely it couldn't be as simple in its composition as Miescher's acid! The compound that dictated our whole complex physical structure must surely be a complex molecule. Attention centered for a time on protein, made of twenty different subunits. But in the end, Miescher was right. People today are fond of saying that we couldn't understand the linear DNA code until the advent of computer binary codes got us thinking about how much information can be packed into the *sequence* of simple, repeated elements. But even in Miescher's day, they had the simple, repeated dits and dahs of Morse code! The only limit is that if you have fewer letters in your alphabet, it takes a bit longer to spell out words.

And DNA is a very long polymer indeed. Even a tiny virus particle, with only a few genes, looks like an explosion in a spaghetti factory when it's broken open. Your own cells each contain about two meters' worth of DNA, and even though it is broken into forty-six pieces, one for each chromosome, it is still a problem for the cell to get all this material neatly divided during mitosis. To keep it compact (as well as to regulate its activity), the DNA in eukaryotic cells is wound around tiny particles made of very basic proteins called *histones.* Acids (like DNA) and

CAREFUL WHERE YOU STEP—WE'RE UNRAVELLING DNA!

bases (like the histones) bind together by their opposite charges, and a large enough quantity of any other charged substance (such as table salt) will interfere with this binding and allow the DNA to spring free.

What happens when a long molecule like DNA is suddenly free from its histones? Any solution of very long molecules has a number of striking changes in its physical properties. One thing that is affected is the way the solution flows: the long molecules resist flow enough that the solution becomes viscous. This viscosity can be used as a rough indication of how long the molecules are.

Water, however, is very good at dissolving charged molecules. Other solvents, such as alcohols, are not so good an environment for a charged molecule, and in alcohol such molecules will tend to clump together (to aggregate). You will be able to see DNA in this experiment because when enough of the double helixes are aggregated they form a cable like a strand of yarn spun from many wool fibers.

DNA Extraction

In this exercise you will extract DNA from cells and partially purify it. Most of its unusual physical properties result from its being such a long, thin molecule. (To get a rough idea of *how* long and thin, imagine a piece of spaghetti long enough to wrap around the world.) But any enzyme that can cut it up will rapidly change these properties—and your skin is full of such enzymes! So use only specially cleaned glassware and don't touch any part of anything that's going to go into your solutions.

I. The first step is the isolation of cell nuclei. Your instructor will do this by cutting up some calf thymus—a tissue full of cells with large nuclei. The thymus is in a buffered solution of sucrose which is isotonic for nuclei. Next, the cells are broken open to release the nuclei by using a homogenizer. In this device, shown at right, a close-fitting glass ball is moved up and down in a heavy glass tube, and as the cells are forced past the ball, they are ruptured and their nuclei are released. (There are other methods that may be used instead, such as running the thymus through a blender.)

The resulting mess is strained through cheesecloth to remove any gristle or clumps, and the suspension is then given a short, low-speed spin in a centrifuge, causing the dense nuclei to settle to the bottom and other cell debris to remain in suspension. The pellet of nuclei is now resuspended in the buffered sucrose and distributed to you.

II. The second step is lysing the nuclei. But first, put a drop of nuclei on a microscope slide, add a drop of aceto-orcein and a cover slip, and examine the slide. How pure does the suspension look? Note the appearance of the nuclei. Now,

1. Take 40 drops of nuclear suspension (well mixed) in a small test tube.
2. Add 20 drops of EDTA solution, mix gently, and let stand 5 minutes. (The EDTA, a common food additive, removes magnesium and calcium, and this begins to weaken the nuclear membrane.)
3. Add 6 drops of SDS solution, mix *very* gently, and let stand 1 minute. (SDS, sodium dodecyl sulfate, is a detergent—chemically pure but not unlike ordinary detergents—and just as detergents make grease soluble, so SDS dissolves the lipids in the nuclear membrane.)

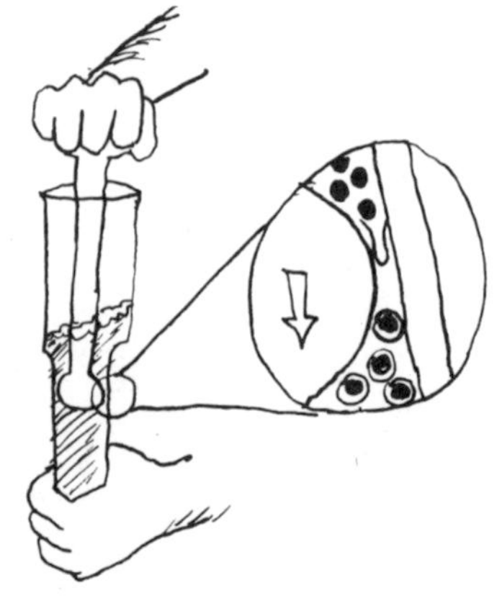

At this point, gently take a sample from your tube for microscopic examination. Use stain as before. How has the picture changed? What you are seeing is *chromatin,* our term for what Miescher called nuclein. From now on, do all your pipetting very slowly, and avoid any harsh or sudden movement of the test tube. The chromatin is rather fragile, and as you do the next few steps it will become more so. For now you begin to remove the protein.

III. Salting off the protein involves providing lots of ions to interfere with the ionic bonding between protein and DNA:

1. Slowly add 6 drops of 2 *M* NaCl to your tube, pausing after every drop, gently mixing. (The mixing should be a slow, circular movement of a clean pipette tip.)

Your protein, now removed from the DNA, will change the appearance of the contents of the tube—how?

IV. All that remains is the separation of the DNA from the protein solution. A simple step that precipitates one substance and not the other is best, and there are many such steps available. We will use one of the simplest, alcohol precipitation.

Here is a step where steady hands are an asset, for the object is going to be to layer the ethanol on *top* of the solution in the tube, rather than mixing it in. But ethanol, unlike oil, will mix with water; you are simply taking advantage of its lower density to keep the two separated.

1. First, get yourself a pipette and some 100% (absolute) ethanol (not denatured alcohol). Then sit down—you simply can't do this standing up.

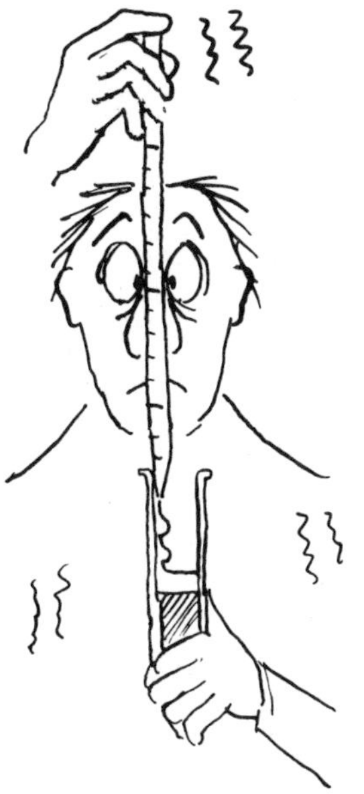

2. Hold your tube of salted chromatin so that you see it against a dark background. Fill your pipette (about 2 mL, but it need not be exact) with ethanol, brace your arms firmly on the table, and now . . .
3. The Moment of Truth. Take a deep breath, let it out slowly, and trickle the ethanol down the side of the tube, slowly at first until you see a layer form on top of the chromatin solution, then faster—just don't run it in so fast that you agitate the layer.

   From now on, don't jolt or shake the tube, or else the layers will mix and you'll be left with nothing but salty vodka.
4. Now, take a very long, thin, and scrupulously clean wooden applicator stick (without the cotton) or a clean glass rod or disposable pipette whose tip has been formed into a small hook or knob by holding it briefly in a flame. This will be the probe with which you will begin spooling out the DNA. First, slide it into the tube far enough to dip below the boundary between your two layers, where you probably already see a whitish precipitate forming; then use a gentle, circular motion as if you were trying to scoop something off the walls of the test tube to begin the spooling process. Dip slightly down into the DNA layer and then up into the alcohol, and repeat. You should soon see strands of DNA winding out on your rod; if not, dip in a little deeper and pull up a little higher until you do. Once the spooling starts, you can be a little less gentle.

Keep on spooling until you have a large clump of DNA on your probe. Would you have guessed that there could be this much in such a small amount of tissue? If the DNA is of very high purity, it will be pure white or, to be more accurate, somewhat iridescent, like an opal. Take the time to examine it closely in different lighting conditions; most people are struck by the fact that this important molecule is also surprisingly beautiful.

If you want to save your DNA, transfer the pipette, or the tip of it, to a tube full of ethanol and seal it well.

If time permits, spool out some of the DNA and dry it down by blowing on it to remove most of the alcohol. Then gently stir it into a little of your 2 *M* NaCl solution in a clean test tube. It may take a little while to get it all dissolved. How does the liquid in the tube seem to change as the DNA is dissolved in it? And why do you suppose the suspension of nuclei didn't become like this immediately after lysis with SDS?

Get a pipette—a 0.1-mL or 0.2-mL pipette will work best, but you can use a disposable one if you draw out a fine tip on it in a flame. If you are using a pipette with no markings on it, make a mark with a felt pen or piece of tape at any position along the pipette you wish. Now fill the pipette up to your mark (or up to a particular mark on a graduated pipette) with water, and then, holding the pipette vertical, release your finger from the pipette end and let the water flow back into the tube, while timing the process. Note down how long it took; this time is a rough measure of the viscosity of the water. Repeat this (using the same pipette filled up to the same mark) with your DNA solution. The ratio of the time it took the DNA to flow out and the time it took the water is a measure of the viscosity of the DNA solution.

Then—experiment. You were told that the ability of a solution to flow (its viscosity) was proportional to the length of the molecules in it (all other things being equal). Can you test this out with DNA? There are a number of things that can chop DNA into smaller (shorter) pieces. One of them was mentioned earlier in this exercise. Another is something that may surprise you—the reason you have been told to be gentle so often in this exercise is that DNA is fragile. You can literally break it (when it is in a pure solution like this) by stirring too vigourously. You may also wonder, Does the stuff change when it is heated? Can exposure to sunlight break it down? What about acids or bases?

Set up an experiment of your own design to test whether a given treatment will change the viscosity of your DNA solution. Remember that if you are going to add any solutions to the DNA solution, you need a control tube diluted with the same amount of salt solution, because diluting the DNA solution with anything will also change its viscosity. Also, make up enough DNA solution to begin with to see you all the way through the process; you need a standard concentration for all your experiments.

If one of your treatments involves a process

(say, shaking or pipetting vigorously), you might test the viscosity after various time intervals and represent your data as viscosity (plotted on the Y axis) and length of treatment (plotted on the X axis). Does the viscosity change steadily, or is there a sudden change at one critical time?

How pure is the DNA you have extracted? One of the simplest ways to tell this is a spectrophotometric test that depends on the ultraviolet light absorbed by the two main things in your sample, DNA and proteins. DNA, because of the purine and pyrimidine rings, absorbs light very strongly at a wavelength of 260 nanometers, while proteins, due to phenylalanine and tyrosine, absorb best at 280 nanometers.

If you have access to a UV-visible spectrophotometer, take some of the DNA you dissolved for the viscosity test (the fragmented, low-viscosity stuff will work well) and put it into a cuvette. For UV work, you will need special cuvettes, since ordinary glass absorbs ultraviolet light. Blank the instrument at 260 with the pure saline solution (see "The Tools of the Trade" for the directions) and take a reading of your DNA sample. If necessary, dilute it (as you did the colored solution in "The Tools of the Trade") until your absorbance reading is below 1.5 units. If you do dilutions, keep careful records of exactly how you diluted the sample.

Then reset the wavelength to 280 nanometers and blank again with the saline (if you must use the same cuvette, rinse it thoroughly first); then determine the absorbance at this wavelength of your dissolved DNA.

Record both values and then determine their ratio—absorbance at 260 nm divided by absorbance at 280 nm. If your absorbance ratio is 1.8, your DNA is pure enough for genetic engineering! Even a ratio of 1.4 is relatively protein-free and is what you should expect from this procedure.

Because of the fact that the absorbance of light is proportional to the concentration of the material absorbing, you can estimate the amount of DNA in your preparation. To do this really accurately, of course, you would make a standard curve with pure DNA at different known concentrations and compare your reading on your own batch of DNA to the readings on the standards. But such standard curves work out consistently enough that you can make a rough estimate from the fact that generally, in a standard 1-cm cuvette, a 50 microgram per milliliter solution of DNA will have an absorbance of 1.0 units.

So, the unpromising-looking goop on your pipette is the same stuff that is responsible for your red hair, your AB blood, your turned-up nose, and your tendency to asthma. Take a good look at this strangely wrapped package of inheritance; it is a legacy from ancestors you know—and from more remote ones from whom you'd run on sight.

It is, in a way, as simple as it looks: just four kinds of beads on a double string. Much too simple, this four-letter alphabet. But take a look around you at this crowded riot of a world—ants and asters, beagles and blue crabs—and think of how much those four letters can spell.

Questions:

1. Could you use the information from this to determine the amount of DNA per cell in calf thymus? What additional information would you need?
2. Why do you suppose you had to be careful in handling the preparation after lysing the nuclei?
3. There are a number of different types of protein in chromatin, and they can be removed by differing amounts of salt. Could you adapt the procedure given here to make a way to determine which of the proteins is primarily responsible for the close packing of DNA in chromatin?

Chloroplasts

All over the world the quietest living things spread out their surfaces to the sun and harvest its energy. From them, from the energy trapped in the molecules they make, all the rest of us ultimately get our energy. For only plants can harvest solar energy and use it to make carbohydrates. We are dependent on them for food and fuel (remember that coal and oil are the residue of ancient plants).

Superficially, the chemical reaction by which carbohydrates are formed is simple:

$$12\,H_2O \xrightarrow{\text{light}} 24\,H + 6\,O_2\uparrow$$

$$6\,CO_2 \xrightarrow{\quad 24\,H \quad} C_6H_{12}O_6 + 6\,H_2O$$

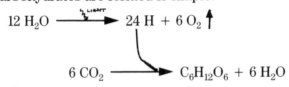

In the presence of light, water is split and oxygen released . . .

. . . the hydrogen atoms combine with carbon dioxide to form sugar.

So why, you ask, can't you make sugar water by leaving bottles of carbonated water (H_2O + CO_2) out in the sun for a while? If you do the experiment, all you get is hot club soda (or maybe burst bottles). The reason is that the equation above only describes the net result of photosynthesis. It is like saying that the way a television set works is that you plug it in and a picture appears on the screen.

The real workings of photosynthesis involve light-trapping pigments, particularly chlorophyll, and a sort of bucket brigade of small molecules which transport the electrons removed from water (the protons follow along to form a complete hydrogen atom at the end) to an intermediate receptor. This is the "light reaction," so-called because is requires light. The top line of the reaction above shows it. The "dark reaction" (bottom line above) does not require light; it consists of a series of enzymatic reactions which use CO_2 and the above-mentioned intermediate receptor to make carbohydrates.

It is possible to step into the bucket brigade of the light reaction with an artificial electron receptor, such as a dye that changes color when reduced. This interference is termed the Hill reaction, and the dye (there are several possible ones) a Hill reagent. It allows us to have a simple visual test for the light reaction. As another approach to studying photosynthesis, we can isolate some of the light-absorbing pigments by extracting them from the cell and separating them by paper chromatography. But first we will take a look at the package all this photosynthetic machinery comes in—the chloroplast. Like so many important systems in the cell, the photosynthetic system is organized and compartmentalized in a system of membranes.

Observing Chloroplasts

Get your microscope set up properly, and then get a leaf or two of the common aquarium plant *Anacharis* (sometimes called *Elodea*) and put the leaves in a drop of tap water on a slide. Hold the leaf at one end with a pair of forceps and slice it several times with a razor blade or scalpel held almost parallel to the surface. This will produce a long, wedge-shaped bit (or bits) of leaf in at least some of which you can distinguish several cell types. When you have cut several bits of leaf, add a coverslip and examine the result, scanning first on low power to find a relatively transparent part of the sample. Then examine your transparent part with the 40 × objective. The most obvious feature is the cell

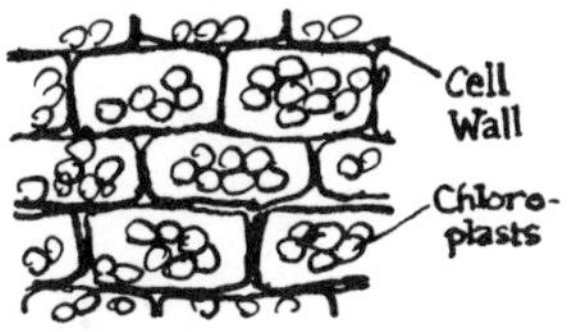

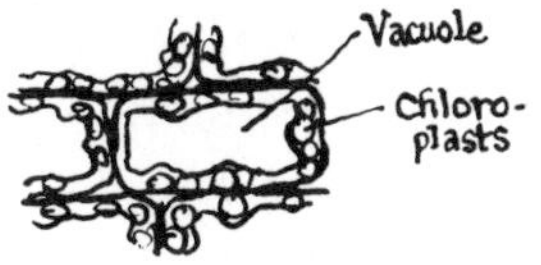

wall, forming outlines around each cell with a pattern like a brick wall. Each rectangular box encloses one cell. Within each cell are many oval, vivid green bodies; these are the chloroplasts. Some cells have a large, clear vacuole in the middle (for storing products of metabolism) and chloroplasts around the outside. These cells, if they are healthy and have been in the light, will show a movement of their chloroplasts within the cytoplasm. This cytoplasmic streaming, or *cyclosis*, is thought to aid the exchange of gas and nutrients by keeping the cytoplasm stirred up.

Questions:

1. How many different types of cells could you see? Describe and draw each.
2. For each type of cell, what was the average number of chloroplasts? What do you suppose differences in this between cell types might reflect?
3. Did anything seem to increase the rate of streaming? To decrease it? Was it smooth, or did the streaming start and stop?

Photosynthetic Pigments

Note: From this point on, there must be no open flames in the laboratory.

Your instructor will grind up 10 deveined spinach leaves in 200 mL 1:3 petroleum ether:methanol to extract most of the pigments, filter out large particles, and then shake 100 mL of the filtrate with 500 mL of 10% NaCl to remove proteins. Finally 50 mL petroleum ether will be added, the mixture shaken, and the top layer removed. This is the material you will chromatograph. *It must be kept in dim, incandescent light; direct or fluorescent light will bleach it.*

Get a strip of chromatography paper and draw a pencil line (*not* ink) across it about 2 cm from the bottom. Keep your fingers off the surface of the paper, handling it only by the edges. Take the smallest capillary tube you can get (or a disposable Pasteur pipette drawn out in a burner flame to a thin capillary) and dip it into your ether fraction, which should be very dark green. Touch it to the paper strip in the center, on the line you drew, so that the pigment is absorbed into the paper. Blow on it as you do this, to speed up drying and to avoid spreading the spot. Repeat the application at least 10 times; you want a very large deposit of the pigments. Let the spot dry each time before applying more. When you are done, suspend the paper as shown in a jar with enough solvent (15:1 petroleum ether:acetone) to reach the bottom of the paper, but not so much as to cover the line of pigment. If the strip curls and seems likely to touch the side of the jar, put a crease in it lengthwise (use clean forceps, not fingers). Put the jar away in a dark place after covering it with foil.

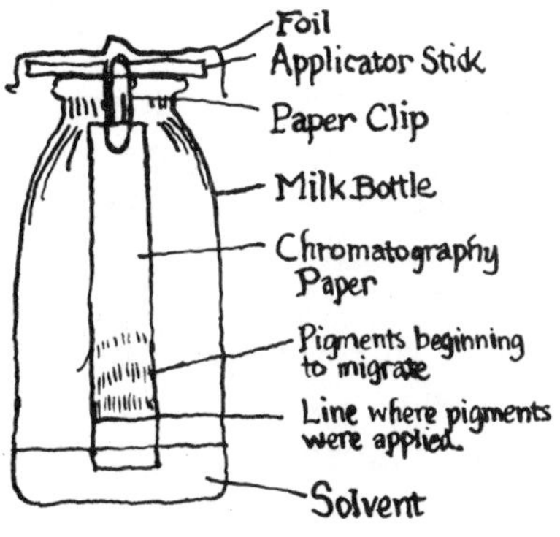

In 20–30 minutes, look at the chromatogram. If the solvent has reached the top few centimeters of the paper, take the paper out and mark the highest point the solvent reached with a pencil line. (If it hasn't reached the upper part of the paper, leave the chromatogram for a longer time.) Look for bands of pigment on the paper and mark them with a pencil outline. Note the color of each band.

The examination of the chromatogram will be easiest when it is dry. Use dim light at first, only bright light to observe the faintest bands. After all the bands are marked, if an ultraviolet lamp is available, use it to cause the pigments to fluoresce. Are the fluorescent colors the same as the colors seen by incandescent light?

The pigments you see may include the following (colors are those seen in incandescent light): chlorophyll *A* (blue-green), chlorophyll *B* (yellow-green), xanthophyll (yellow), carotene (yellow-orange), anthocyanin (red to purple). The first two are by far the major pigments in green leaves. It is only when they break down, as in fall leaves, that the other pigments become visible.

You can use this technique (which is related to the first use ever made of chromatography by the Russian biologist Tswett) on any plant material at all. How do the pigments differ in colored plant leaves such as those of Coleus? Do flower petals contain any of the photosynthetic pigments? Time and materials permitting, try to answer these questions using the techniques above. If a pigment has a different color under visible or fluorescent light, or a different R_f value (see below), it is a different type of molecule.

Questions:

1. The R_f value is an expression of the distance a given molecule travels in a chromatogram, expressed as a fraction of the total distance the solvent moves. Measure the distance from the pencil line you put the pigment on to the line you drew at the top edge of the solvent when you took the chromatogram out. Call this *A*. Now, for each pigment, measure the distance from the starting line to the middle of the pigment band; call this *B*. The fraction *B*/*A* is solved to obtain the R_f of that pigment. R_f values are decimals, and the highest possible is 1.0—a pigment right at the top. For the same solvent system, these R_f values ought to be the same from one chromatogram to another; so (*a*) give your R_f values for each pigment, and (*b*) compare them to those from the rest of the class.

2. You had to keep your pigments in dim light to prevent bleaching, but plants certainly don't bleach in the light. How can you reconcile this? (Note that even dead plants don't bleach out in the light for a long time.)

The Hill Reaction

Your instructor will have prepared a crude chloroplast fraction from spinach leaves by grinding up the leaves and spinning out the debris at low speed in a centrifuge, then spinning out the chloroplasts at a higher speed. You will use a Hill reagent to show that these isolated chloroplasts can still carry out the light reaction. The dye is colored, but when it is reduced by the light reaction it becomes colorless.

First, take two test tubes with 1 mL of chloroplasts in each. To one add 3 drops of distilled water; to the other, 3 drops of Hill reagent. Mix gently and expose to a lamp or to direct sunlight. Observe the length of time needed to bleach the tube with dye to the color of the one without. If you are using lamps, be sure that they do not heat up your tubes to the point that they feel hot to the touch.

Since you are by now a properly skeptical scientist, you will want to add 3 drops of Hill reagent to a tube containing 1 mL of *water* and expose that to the light as before.

You will also wish to wrap a tube in aluminum foil and add 1 mL of chloroplasts and 3 drops of Hill reagent to that, and also set up unwrapped tubes with the same contents in different degrees of lighting.

Put a tube containing 1 mL of chloroplasts (but no Hill reagent) in a boiling water bath for a minute or two. Cool it off and then add 3 drops of Hill reagent. Note the result.

If time permits, design an experiment to show whether the Hill reagent must be present at the time of illumination in order to be reduced. Carry out the experiment you have designed.

Questions:

1. How long did it take to bleach the Hill reagent in maximum light? In dimmer light?
2. What happened in the tube with chloroplasts and water?
3. What did you learn from the tube with water and Hill reagent?
4. What was the effect of boiling the chloroplasts? What does this tell you about the type of molecules involved in the light reaction?
5. What was the experiment you designed? Explain the result. (If you did not have time, then design the experiment now and write down how you would do it and why.)

Postscript

It is thought that each light-absorbing unit in a chloroplast is a huge, antenna-like battery of several hundred chlorophyll molecules; they all pass their captured energy on to a single central one, which feeds it into the light reaction. In spite of this, it has been calculated that a field of corn uses only 1 or 2 percent of the solar energy falling on it, the rest being lost. That seems to be fine for the corn, but in this energy-saving age it is a little shocking to us consumers!

The Cell Surface

The inside of a cell, as you know, is a bewildering array of molecules both great and small, which are engaged in the great dance whose steps may be seen on a metabolic pathways chart. Outside the cell (be it protozoan or metazoan) lurks a much less ordered world, quite inhospitable to the complex molecules upon which the life of the cell depends. Between the inside and the outside is the *membrane,* a film of lipid of the consistency of olive oil, a mere tenth of a nanometer (10^{-9} m) thick. Floating in this miniature oil slick are proteins of many kinds, which are mainly concerned with regulating the flow of materials between the inside of the cell and the world outside. Some proteins, at least in metazoans, are also involved in communication and contact between cells. Many of the surface proteins and some lipids are coated with sugar in the form of complex oligosaccharides. This whole array with which the cell confronts the world is called, with deceptive simplicity, the *cell surface.*

One remarkable thing about the cell surface is that it is not a perfect barrier to everything. It is *semipermeable*: it will retain large molecules (proteins, etc.) and small, charged ones, but water can pass freely through it. Think about the consequence of that for a moment. When you boil sausages, they burst; they are relatively dry inside and so they take up water, but the casings are not permeable to sausage, so the total contents of the casing increase to the bursting point. Cells behave in much the same way: if they are put in a solution with fewer dissolved particles (and therefore more water) than the cell has—that is, a *hypotonic* solution—a cell will take up water and eventually burst. A solution with more dissolved particles (and therefore less water) will cause a cell to give up water and shrivel up; this sort of solution is *hypertonic.* A solution that has a number of particles similar to the number inside the cell produces no changes and is called *isotonic.* (The particles can be virtually anything that cannot freely cross the membrane, such as proteins, salts, and so forth.)

Protozoa have very elaborate means of dealing with this, mainly consisting of a membranous water pump in each organism. Metazoan cells usually do not face changes in the particle content of their environment, since the metazoa have generally developed a specialized group of cells, a kidney, to regulate the internal fluid composition of the whole organism. But when the cells of a metazoan are isolated from its body and subjected to a solution that is hypotonic for that organism, the cells swell so uncontrollably that they finally explode! This alarming phenomenon is technically termed *lysis.*

The cell surface is studded with various sugars and proteins, many of them serving unknown purposes. Some of them aid cells in holding on to their neighbors and in recognizing the cells that belong nearby and those that do not. It may be that a good deal of the blueprint for the architecture of tissues and organs lies on the cell surface, like the array of holes on a Tinkertoy block.

One surface substance known since the first years of this century is a sugar chain called the ABO System Blood Group Substance. Like all biological molecules, it has genetic variants; and in this case, most people also make *antibodies* against substances different from their own. (An antibody is a protein made in response to a foreign substance—the *antigen*—having the ability to bind to this antigen, initiating a variety of processes all aimed at the swift removal of the offending matter.)

In the case of the ABO system, the antigen is a sugar chain attached to a lipid molecule in the membrane of red blood cells. It has a stalk, the *H substance,* and either the A or the B substance, or both, may be attached to the stalk. The consequences are summarized this way:

If you are blood type . . .	Your blood group substance looks like . . .	And you make antibodies to . . .
O		A and B substances ("universal donor")
A		B substance
B		A substance
AB		neither A nor B ("universal recipient")

It was an early discovery that the small quantity of the donor's antibodies transferred in a blood transfusion doesn't affect the recipient; but the donor's red cells, if the recipient makes antibodies against them, get clumped together and do things much too horrible to print here to the recipient's circulation. For this reason blood is carefully tested with antibodies ("typed") to identify its antigens before it is used for transfusions; these tests work by checking whether, say, an anti-B type antibody is capable of clumping together the red cells being tested while they are still safely outside the body. If they can, then the red cells are type B or AB and would not be given to a person with type A or O blood. Besides the ABO system, there are dozens of other antigen systems on red cells—for example, the Rh antigens, which are so famous because of the troubles that can be caused by an Rh incompatibility between mother and fetus. Not all antigens of this sort are on red cells, nor are all of them made of sugar chains—but for very few of them have we any idea what they are doing there at all! In some sense, though, they are certainly markers of our genetic individuality.

A hint of what some cell-surface antigens may be doing can be gained from relatively simple organisms, which don't mind being pulled apart. Remember the sponges? The cellular anarchists? They can be pulled cell-from-cell and still get back together again. This in itself argues for some sort of cell recognition; but suppose the cells are just sticky and get together that way? If there really is something on the surface of cells that says, in effect, "I'm me and you're you," some marker of individuality, then separated sponge cells should not only find their own kind but shun strangers! Suppose we mix two kinds of sponge cells together—will they sort themselves out? We will see.

Reconstitution of Sponges—I

We will use two species of sponge, one red-orange and one yellow. Before you start, take a good look at each so that you can recognize the colors.

With a scalpel, cut out a bit of sponge roughly 1 cm cubed. Cut this into smaller pieces at your desk and put them in the middle of a square of bolting silk. Obtain 20 mL of sea water in a wide-mouthed jar or beaker, and, folding up the silk like a bag, lower the sponge fragments into the water. Now, by twisting the top of the bag, squeeze the sponge through the silk. You will be able to see color in the water as the cells come through.

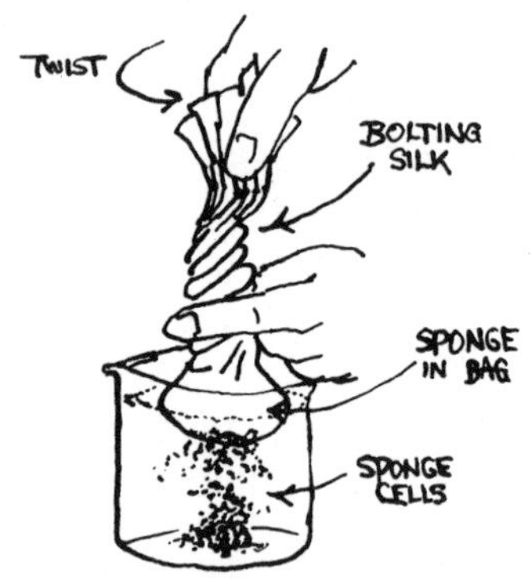

Squeeze until most of the bulk of the sponge is gone; some will remain, but you can get most of the cells out. Label this jar with the color of the sponge you used, and do the same straining procedure over again with the other kind of sponge. Be sure to wash out your silk thoroughly, first.

You are now ready to set up experiments. First, as controls, you need to let each type reaggregate independently, so into a petri dish marked 1 put 2 mL of your first suspension of sponge cells (gently stir the beaker up before each removal of cells from it) and immediately add 8 mL of sea water. Dish 2, naturally, gets the same dilution of your second suspension of sponge cells (the other color). Dish 3, your experiment, gets 2 mL of *each* suspension and 6 mL of sea water. You will note that you still have 16 mL of each suspension; now you may design a few experiments of your own. For example: (1) Does temperature affect the reaggregation process? Try refrigerators, warm rooms. (2) Does concentration affect it? Try more and less dilute suspensions. (3) Does mixing affect it? Shake one plate gently at regular intervals.

Try anything that comes to mind, but remember to use controls and to avoid conditions that simply kill the cells. (Inspect them under the microscope if you are not sure; dead cells disintegrate.) Incubate all dishes (except temperature experiments) at room temperature, undisturbed, and go on to the rest of this exercise. You will inspect your plates later.

Osmotic Pressure

You will be given some fresh blood and a variety of concentrations of salt solution. Using these, and your microscope, you will determine what salt concentration is isotonic for these cells.

Take as many vials as there are solutions and label each with a different concentration. Put about 25 drops of the 1% salt solution into the vial labeled 1%, and so on. Cork the vials. Set up your microscope with care, and with the lamp as dim as is convenient; then add a drop of blood to the first vial. Mix gently, wait one minute, and then examine a drop of this solution under the microscope. If you have a partner, or if you work

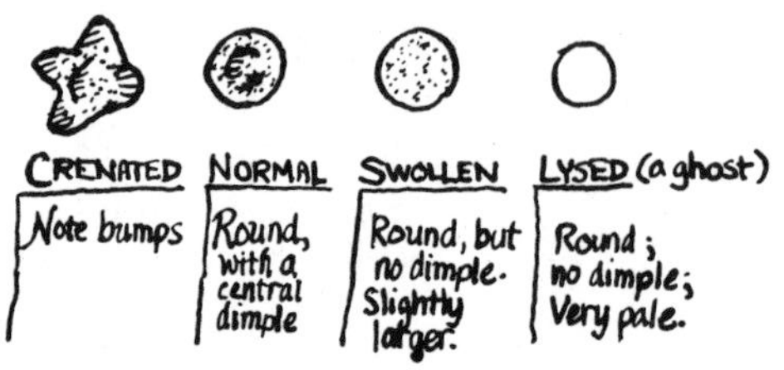

slowly, seal the coverslip with Vaseline. Note the shape of the cells and record this next to the percent salt you used. Cover the vials after you add the cells to them, and keep them; you may want to make comparison slides from them later, since it is difficult to make accurate estimations of size using the microscope without special attachments.

After you have looked at cells in all the concentrations of salt, try to decide which is the "normal," i.e., isotonic, concentration. Also, try looking through the vials at a page of rather fine print or graph paper. Is it possible to see clearly through any vials? Note down which ones you can see through.

Questions:

1. What is your estimate of normal saline's percent concentration? Is there some ambiguity in it? What could account for this ambiguity?
2. Is there any correlation between your ability to read through a vial and the cells' condition? Why?

Blood Cell Antigens

This is an experiment that is often done, but is seldom done right. The author repeated it in numerous exercises as a student and often turned out to be AB, which was a great shock to his parents, who are both type O! Care must be taken to ensure against false positives, so attend closely to the following steps.

1. The first step is getting a couple of *clean* slides—any dust or finger grease will ruin things. Clean them with alcohol and wipe dry.
2. Make one 2-cm circle in wax pencil on the slide for each antigen you will be testing; label each circle (*A*, *B*, and so forth). This makes a well.
3. Dispense a drop of reagent ("anti-*A*," and so forth) in the appropriate well; prepare several wells this way *before* drawing blood (this is to prevent contaminating reagents with blood).
4. *Now comes the hard part.* Swab the end of a finger with alcohol; let it dry. Get your lab partner to puncture the skin with a fresh, sterile lancet; it is much easier than doing it yourself.
5. Discard the first drop of blood. Add each succeeding drop to one of the wells, one drop to each.
6. Immediately mix each well's contents with a clean toothpick (use a fresh one for each well).
7. Gently rock the slides from side to side to mix; do this for 2–3 minutes.
8. Now tilt the slides toward you so most of the liquid drains to the bottom of the well. Score each well as indicated below:

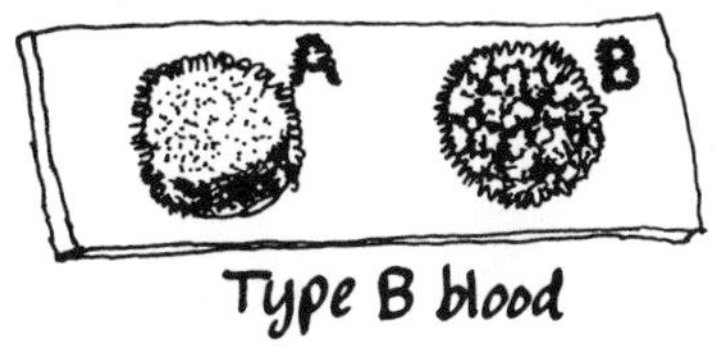

Well *A* is *negative.* There is a thin, uniform sheet of red cells in the upper part of the well.

Well *B* is *positive.* There is an intricate network of red cells in the upper part of the well.

Compile your results for each blood group; write your blood type. (If you want to keep the slides, tilt them and blot off the excess liquid, allow them to dry overnight, and seal with Scotch tape.)

Questions:

1. Basing your knowledge on the table given earlier in this exercise, draw your ABO substance.
2. The percentage of the population in the U.S.A. having a certain blood type is given (for some types) below:

ABO System	Rh System	MN System	Ss System
O: 44%	Rh_D (Rh+): 85%	M: 27%	S: 11%
A: 42%	Rh_d (Rh−): 15%	N: 24%	s: 45%
B: 10%		MN: 50%	S: 44%
AB: 4%			

How does your class compare to these figures? How many people are a perfect match for all the substances tested?

Reconstitution of Sponges—II

Examine the dishes you set aside earlier. Your dissecting microscope will be helpful. For each dish, record (*a*) presence or absence of clumps; (*b*) approximate diameter of clumps; (*c*) color of clumps (this requires reflected, rather than transmitted, light). Compare experiments with the class.

Questions:

1. Did you observe reaggregation under any conditions? If so, which?
2. How big did the aggregates get in each case? Does this tell you anything about the number of cells involved?
3. Did you observe sorting-out of the different types of sponge? Under what conditions did you observe this?
4. From your own experiments, and from the class's, can you make a list of conditions required for reaggregation? For sorting-out? From the list, can you formulate a general rule? How might you test this rule?

Mitochondria

Buried in all your tissues, and tucked into every other type of true cell we know of, are the famous sausage-shaped organelles called *mitochondria.* Although they have been called the "powerhouses of the cell," this may create some false impressions. Many people take this to mean that they are *sources* of the cell's energy. But of course they cannot be, any more than a powerhouse can be a source of energy. Something else—coal or oil, the sun or a waterfall, or nuclear fission—is the source of the energy. The only job a powerhouse has is to make that energy available in a useful, standard form, such as AC electrical current.

In the case of mitochondria, the energy is contained in organic molecules such as carbohydrates or fatty acids, and the mitochondria are concerned with splitting the bonds in those molecules in such a way as to produce a useful, standard form of high-energy molecule, in this case ATP. The standard way the cell's energy-requiring processes are made to go is by coupling them to the breakdown of this ATP (the products are recycled). So why, you ask, don't we just eat ATP and cut out all this sugar and fat business? Apart from the fact that it is more fun our way, it wouldn't work. ATP must be made by each cell for its own use; it cannot (as far as we know) be shared among cells. To this degree, you see, you are still a community of separate organisms.

This means, of course, that mitochondria are pretty common things. All eukaryotic cells have them—repeat, all. A favorite trick question of the makers of biology tests is constructed to try to get you to say that whereas animal cells have mitochondria, plant cells have chloroplasts instead. In fact, plants as well as animals have mitochondria, and the source of the ones you will use in this lab is a flower!

There are two stages in the process carried out by mitochondria. In the first, a carbohydrate is cleaved into carbon dioxide and hydrogen; in the second, that hydrogen is combined with oxygen from the atmosphere to make water. The overall scheme looks like this:

$$C_6H_{12}O_6 \longrightarrow 6CO_2 + 24\,H$$
$$6\,O_2 + 24\,H \longrightarrow 12\,H_2O$$

This ought to look familiar. Yes, it is the scheme of photosynthesis in chloroplasts, backwards. You could say that what mitochondria do is approximately the opposite of what chloroplasts do.

To be accurate about it, however, the mitochondria don't carry out the whole process as written; the beginning, called *glycolysis,* goes as far as splitting the 6-carbon sugar into two 3-carbon fragments before the mitochondria are involved at all; this happens in the cytoplasm.

Also, if you remember (or looked up) the overall reaction of photosynthesis, you notice that the wavy line labeled "light" is missing here. Instead of glowing in the dark, your mitochondria immediately convert the released energy in the last step into the chemical bonds in ATP.

There are procedures given here that will allow you to investigate two of the major events in mitochondrial function. Each is rather time-consuming, so you may be instructed to select one or the other, or else extend this lab over two periods.

The first event is the removal of hydrogen from a carbohydrate. Shifting an entire "H," as written in the equation above, involves moving a portion and an electron, and in the chemist's language, anything that has an electron removed from it is said to be *oxidized*, and anything that has an electron added to it is said to be *reduced.* Since, in most biological oxidations and reductions, the proton comes along for the ride, it is pretty easy to tell that, for instance, when FAD is converted to $FADH_2$, it is being reduced. (Notice that something else must have been oxidized in the same process, since the H's had to come from somewhere.)

In our case, the source of the H's is a 4-carbon acid called *succinic acid* or *succinate*, in its ionized form. It is a key product of the Krebs Cycle, which is the pathway in mitochondria that breaks down dietary carbohydrates (and other things) all the way to carbon dioxide. Succinate is oxidized to fumarate, losing two H's in the process, and the FAD mentioned above is reduced. All of this is mediated by the characteristic mitochondrial enzyme *succinate dehydrogenase.*

Normally, the $FADH_2$ then dumps its H's into a sort of molecular bucket brigade called the Electron Transport System, in which a series of reducible molecules, mostly the colorful cytochromes, hand the electrons, sometimes with their protons and sometimes without, to each other like a pair of hot potatoes until at last they joyfully unite with oxygen to form reduced oxygen, perhaps better known as water.

Small disgression: yes, you really make water from hydrogen and oxygen when you oxidize your blood sugar. And some small desert mammals get along apparently without drinking water at all, just by using this "metabolic water" supply.

Second small digression: yes, joyfully unite. In physical terms, this means with a concomitant release of energy. Very few teachers have the courage to do the classroom demonstration of combining a little oxygen with twice as much hydrogen gas and sending a spark through it—the explosion that results can be a little devastating. That reaction drives many rocket engines. Your mitochondria do it more quietly, to be sure, but not because less energy is released—rather, because they trap it to make ATP.

The reaction of succinate dehydrogenase, then, is one way in which H can be poured into the chain of reactions that ends by making available the energy to produce ATP. But how can you see a reaction like that? By using a dye that undergoes a color change when it is reduced by $FADH_2$, and fortunately there are several. To increase the chances of seeing the reaction, you will also poison the last step of the pathway, the reduction of oxygen; this will allow lots of H to accumulate. Sodium cyanide can do this, but its ability to form the lethal gas hydrogen cyanide if it contacts any acid makes it unpopular with teachers (at least ones who like their students). We will use the equally toxic but nonvolatile sodium azide instead.

The basic scheme will be to measure the activity of the enzyme in purified mitochondria and then determine whether another similar molecule, malonate, can be used as an H donor instead of succinate.

The second procedure allows you to measure the uptake of oxygen by mitochondria and to examine the effects of several metabolic poisons on the process. Instead of doing a chemical assay for oxygen, you will just monitor the changes in the volume of the gas over a suspension of mitochondria. This gas uptake is called *respiration*, and very appropriately so. It is the sole reason that you must breathe in an oxygen supply.

Since mitochondria release carbon dioxide gas, you cannot simply measure changes in total gas pressure; these would partly cancel out. You will use a paper wick soaked in strong potassium hydroxide solution to absorb the carbon dioxide from the air as it is released. The decrease in gas volume will then be a fairly accurate estimate of oxygen uptake.

Of course, the first step is to obtain mitochondria. Perhaps the source—a cauliflower—will surprise you. Remember that even plants have mitochondria. The outermost florets of the cauliflower will be cut off (you may be asked to do this) and ground up, after which the preparation will be centrifuged once at a relatively low speed to remove nuclei and (if your centrifuge allows it) again at a higher speed to sediment the mitochondria. The whole procedure is similar to the one you followed in the chloroplast lab.

Succinate Dehydrogenase

To do this assay correctly, you will need to have good pipetting technique and must take care to put the right thing in the right test tube, and in the order indicated. (You can tell if most of your class is getting it right. The lab will be very, very quiet.) One suggestion: label each tube *first*, and then check its identity each time you add something to it.

The mitochondrial suspension you will be working with can settle out over time; so every time you need to draw some out, mix the tube first and remove the amount you need quickly, while the mitochondria are still suspended.

Get five clean test tubes (more if time and supplies allow you to do some additional investigations) and label them 1–5. Keep the mitochondrial suspension on ice and add it to each tube last of all.

Two of these tubes are experimental tubes: numbers 1 and 5. The other three are *some* of the possible controls.

Add to each tube the amounts of the solutions indicated (all volumes are in milliliters):

Tube No.	Assay Medium	Azide	DCIP	Succinate	Malonate	Mitochondria
1	5.4	0.5	0.5	0.5	—	0.6
2	5.9	0.5	—	0.5	—	0.6
3	5.9	0.5	0.5	—	—	0.6
4	5.9	—	0.5	0.5	—	0.6
5	4.9	0.5	0.5	0.5	0.5	0.6

Mix the tubes well as soon as the mitochondria are in them, and record the time. You will be taking absorbance readings from each tube over the next 30 to 45 minutes, at frequent intervals.

Set the spectrophotometer you will be using to 600 nm wavelength and set its zero adjustment with your blank, tube 2. Then take absorbance readings on each of the other tubes and record the reading beside the time you took it.

Repeat the process above (including the zero adjustment) every five minutes, gently mixing the tubes before each reading. Record the exact time of the reading; then if it was really 6 minutes rather than 5, you can simply mark your graph at the 6-minute point. Time courses like this can seem nerve-wracking to you if you aren't used to them, although they are as common to most biological and medical research as practice sessions are to music. Just remember that you want to keep *strictly* accurate track of the *actual* times at which you took the readings, and they *try* to make them about 5 minutes apart. Don't get flustered or drop tubes in a wild effort to hit the 5-minute mark exactly; that doesn't really matter.

When can you stop? You will notice that the other tubes all have a color (due to the DCIP) that is really what you are measuring; this color gradually bleaches out as the electrons are transferred to it from the mitochondrial enzymes. When tube 1 has either leveled off (same reading three times in a row) or gone below 0.02 absorbance units, there is no point in continuing. If you collect data for 45 minutes and neither of these things has happened, then you have some very lazy mitochondria and you certainly have enough data to tell you something.

Make a graph of absorbance readings (on the Y axis) vs. time in minutes (on the X axis) for tubes 1, 3, 4, and 5. Draw the best line through each set of time points (one line for each tube). Is the best-fitting line straight or a curve?

What effect does malonate have on the reaction? Is it what you expected? Why, or why not? What do you think might be happening? Malonate is a derivative of malic acid, which is found in apples and many other types of fruit. What do you suppose would be the effect on your mitochondria of absorbing a lot of malonate?

(Note to the curious: In the preceding paragraph, I have played a very common alarmist trick on you. Please notice that saying X is a derivative of Y does *not* mean that X may be *found* in Y. To make a derivative, you generally have to do some chemical reactions. Also, lots of things you eat would poison your mitochondria if they contacted them directly. But mitochondria, remember, are contained within cells, and cells have very selectively permeable membranes. See the trick? Once you get the hang of it, you can use this kind of thing to horrify everybody at the dinner table, and if you are very good, you may have a career in popular science writing.)

Oxygen Uptake

The main skill you will need to do this part is the ability to be aware of temperature variations. Since gases expand when hot and contract when cool, the temperature changes induced by your eagerly grabbing one of your tubes with a warm hand could indicate reverse respiration! Set up all your tubes in a draft-free place, handle them only with test tube holders, not with your fingers, and if possible, put them and all your solutions in a large volume of room-temperature water in a shallow pan, or, if one is available, in a water bath. This will help keep temperature shifts to a minimum.

Also, make sure that your mitochondria are only a shallow puddle in the bottom of the tube. If your solution is deeper than the tube is wide, then gas exchange will be somewhat limited.

Finally, when you add the KOH wick, take great care that it doesn't leak some KOH down into the mitochondria. That would kill the reactions. If you even think this has happened, start over with clean tubes rather than sitting waiting for dead mitochondria to respire!

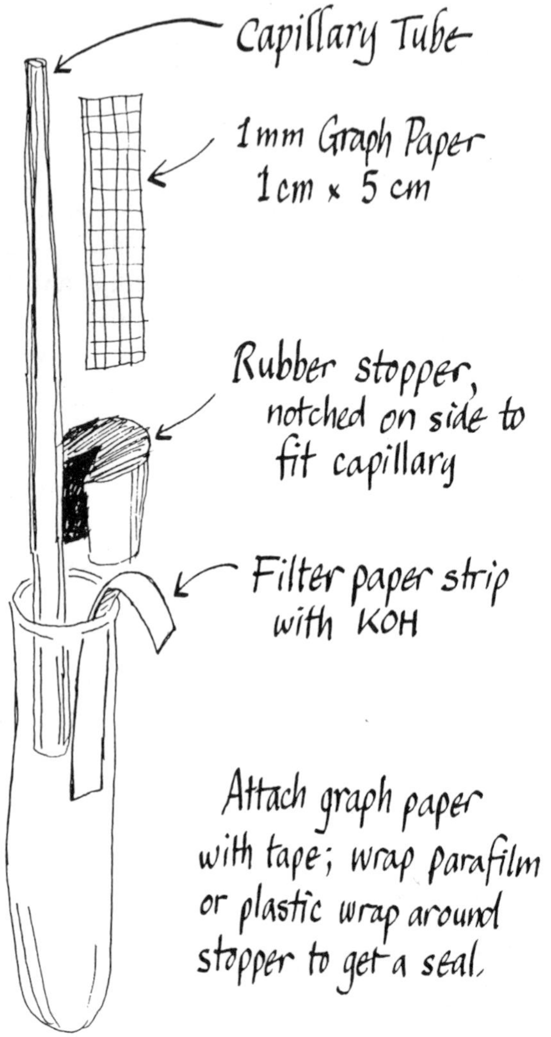

Make the respirometer as shown in the drawing; make five of these. Cut out and pleat the filter paper for the KOH, but don't add the KOH or put it in place until you are ready to run the test.

Set up one respirometer with 1.25 mL of water in it and treat it like the others. It will register any changes in the temperature of the air or water.

To each tube except the one just mentioned, add

0.5 mL assay medium

0.25 mL metabolic poison or assay medium (see below)

0.25 mL succinate

You will have two poisons to test, DNP and cyanide. *Don't mouth pipette or even contact them with your skin.* If you do get any on you, rinse it off immediately. Both of these are very poisonous.

It will be part of your job to figure out, using the scheme above, how to test the effect of these poisons on respiration. You will need to think of suitable controls, for example. Work them out, discuss them with your instructor, and then proceed to make up the tubes as you have decided. Allow them to stand and reach the same temperature for a few minutes; then to each tube add 0.25 mL mitochondrial suspension and place the filter paper dampened with KOH in the neck of the tube; place the cork in it and seal well. Leave the whole assembly in the test tube rack (in the pan of water if you are using one) for 10 minutes to stabilize the temperature.

Then, without touching the tube, add a tiny drop of immersion oil or, better, colored mineral oil to the top of the capillary. Try to make each drop the same size for each tube.

Record the position (in mm) of each drop over the next half-hour or so, or until the drop either stops moving or falls into the tube. Take readings every minute or so, recording the time exactly.

Even the tube with plain water will probably show some movement, but the unpoisoned mitochondrial tubes should show considerably more. Since the capillary tubes are relatively constant diameters inside, the change in position of the oil drop is proportional to the volume of oxygen consumed.

How does cyanide affect the rate of uptake of oxygen in mitochondria? How does dinitrophenol affect it?

It was by studies involving such questions (and basic methods) as these that the surprising way in which respiration works was discovered. They key point was that cyanide blocks electron transport by certain cytochromes, while dinitrophenol transports electrons itself, moving them across the mitochondrial inner membrane. In both cases, however, the poisons would be fatal to a cell, because they both block the production of ATP. This provided a significant clue to the workings of the mitochondrion.

Muscle Contraction

One of the most widespread physiological processes is the one that we typically experience as muscle contraction. Most animals use it in one form or another, and the same process on the molecular level goes on in the flowing cytoplasm of an ameba, even though no muscle fibers can be seen.

Why does muscle act as it does? The fascination with this goes back to the days of the debate between vitalists and mechanists (see "The Cell's Alchemists" for a summary of a later stage in this debate). Is there a mysterious "life force" that magically moves the muscles, or are they somehow little machines that work all by themselves, following ordinary laws? An eighteenth-

century Italian anatomist named Luigi Galvani found that muscles that have been dissected out of the body will still contract if they are given an electrical shock (the legend is that he put an iron skillet of frog's legs on the window sill in the kitchen during a thunderstorm and discovered them kicking like a chorus line). For some people this meant that muscles were machines—electrical machines. For others, it meant that electricity was the mysterious life force, and this led to a tremendous industry of making and selling electrical belts, which were supposed to cure everything from cancer to impotence.

When the microscope was turned on muscle fibers, however, a regular pattern of fine stripes was seen running across the muscle fiber, and the pattern was seen to change as the muscle contracted. The stripes were protein-rich, and in time the proteins were isolated and described. They were of two types, both fibrous proteins (that is, long and narrow molecules). The heavier of the two was the golf-club-shaped *myosin,* while the spherical *g-actin* could stick itself together into long filaments of *F-actin* and so be long and narrow also. It was hypothesized in the 1950s that these molecules somehow interacted by sliding forcibly past each other, and that this accounted for the contractibility of muscle.

There was a small problem with this, however. Pure actin and pure myosin could be combined in a test tube, and far from sliding past each other, they just sat there innocently doing nothing. However, they could be induced to shrink into a compact gel by the addition of a variety of crude cell extracts, and eventually the missing ingredients were discovered. Precisely what they were will be the subject of this lab.

In addition, people about this time began working with preparations of muscle cells that

had their cell membranes removed. The usual method was to soak the fresh muscle in glycerol, which gently dissolves the membranes. Such "glycerinated muscle" still has the typical pattern of stripes across it and will still contract—but not from electrical stimuli, since the electrically sensitive part turns out to be the membrane. Rather, glycerinated muscle responds to the chemical environment it is in. The actin and myosin are directly bathed in any solution you put on the muscle fibers. If the muscle contracts, this will be visible.

Under the microscope, you can see the striations in the muscle, and if you succeed in making it contract, you can determine whether the striations change in any visible way.

The ground rules for experimenting with this system are simple:

1. Do not allow the muscle fibers to dry out at any time. This is very damaging to them, and you cannot use them after they have dried.
2. You can put one solution on the muscle, and, if there is no response, you can put on another solution—but don't do this more than once. The solutions dilute each other.
3. You can't get normal contraction with a big chunk of the fibers, because all their normal connections (muscle sheaths, tendons, etc.) have been removed. They will curl up but not contract normally. So use only the thinnest fiber you can still see—not more than 0.2 mm wide.
4. Handle the fibers with clean dissecting needles, and don't leave the needles in contact with the fibers or solutions more than a few minutes. Metal ions may inhibit the reaction.
5. When you are looking at the fibers under the microscope, take particular care to adjust the substage iris to get good contrast. (This basically means darkening the field a bit by stopping down the iris).

You need controls for any experiment. For this one it will be a positive control, allowing you to see contraction take place.

Simply tease apart some fibers of muscle, put them on a microscope slide in a minimum of glycerol, measure them, and then add a few drops of the solution marked "MAGIC." You should get contraction within about 30 seconds. Measure the muscle again, and examine it under the microscope.

1. How much (what percentage of its starting length) does the muscle contract? How fast does this happen?

2. How do the striations change after contraction? (Compare with an untreated fiber.)

Now you need to figure out what things are needed to make contraction happen—that is, what is in the "MAGIC" tube.

You will have solutions of various salts and organic molecules available, and if you think of something else you would like, you can try that too. Keep careful records of your results.

What materials are needed for contraction, and why?

Dissecting the Cell

In order to study something, it is generally necessary to take it apart—as you have seen with the fetal pig in earlier exercises. This is a fairly straightforward process when the object of your attentions is bigger than a breadbox; but what do you do when it is as small as a single cell? You have seen what a very little slide movement looks like under the medium-power objective of a microscope; imagine trying to use forceps and a scalpel under that objective!

There actually have been people who have done that, on large cells at least. They made fine glass needles and scalpels and discovered that will a little care you can push through the membrane, carve out an organelle or two, and withdraw your needle leaving the cell intact. On the other hand, this is not the kind of thing you would like to do all day, and it still doesn't permit collecting little piles of the structures you want to study.

Other methods were invented for this purpose, and the pioneering work of deDuve in the 1950s led to an approach that is standard today—dissecting a cell without ever looking at it. If this seems a little unlikely, just recall that there are lots of machines designed to sort things out by size. All you need to do is devise such a machine for sorting out the internal parts of cells, and you have the tool you need. As a matter of fact, farmers had been using such a machine for a long time already, and scientists had borrowed it in the nineteenth century. The device is the *centrifuge*—a sort of high-class cream separator. Milk that was allowed to stand in the pails for a while would normally develop a rich top layer of cream which could be skimmed off (hence "skim milk"), and any debris in it would settle to the bottom. If you took the pails and attached them to an axle and spun them around, you got the same result, only faster. The centrifugal force pulled down on the components of the milk as a function of their weight just as gravity did, only the centrifugal force could be increased as desired by simply spinning the buckets faster.

By adapting this design to spin smaller amounts faster than any cream separator, it became possible to separate the components of cells according to their weight. This is, of course, an oversimplification. The density (weight-to-volume ratio) and shape of the components, as well as the density of the liquid they are in, actually determine the rate at which they fall to the bottom of the tube (sediment). And the design of the machine that does this must deal with the great forces acting on the metal of the rotor (the part that holds the tubes) and the heat generated by the rapid spinning. Extremely fast machines (called *ultracentrifuges*) require the rotor to spin in a vacuum chamber to reduce friction, for example. But these are all details by comparison with the brilliant simplicity of the idea of sorting things out by spinning them around.

Cell fractionation (the proper name for this "dissection" of cells) requires that the cells be broken up (that is, their cell membranes ruptured) very gently, and the pieces immediately

put on ice. The reason for this goes back to the original work of deDuve, who studied the *lysosome*, the organelle that contains enzymes capable of turning the whole cell to a shapeless mess in minutes. DeDuve called them "suicide bags." (Why do you have them? Well, the death of cells is sometimes essential, for example in the change from the embryo's flipperlike hands and feet to ones with fingers and toes.) Any ruptured lysosomes (and a few are bound to be there) will ruin your experiment by digesting the structures you are trying to separate. Putting the ruptured cells on ice slows the process down until the structures can be safely spun out of the enzyme-laden soup.

Cells are fractionated for many purposes, ranging from the preparation of commercially usefuly enzymes to the purification of DNA for genetic engineering reseach. The exercise you will do will take you through the purification of a typical cell—a liver parenchymal cell— and then have you test the fractions you obtain for some of their major components: protein, RNA, and DNA. This will give you the basic profile of the way compounds are distributed in cells as well as some experience with typical biochemical tests.

Fractionation Procedure

I. Homogenization

Start with fresh tissue (liver is usual) and carry out the following steps (*Note:* Keep the liver chilled, preferably on ice, during the whole procedure.)

1. Trim off any large pieces of connective tissue, just as you would in preparing liver for cooking.
2. Use sucrose solution to *perfuse* the liver—that is, use a syringe and needle to inject the sucrose solution into the veins and arteries of the tissue until you have washed out as much blood as possible. Completely perfused liver is light brown, not mahogany, in color, and only clear liquid comes out when more sucrose solution is injected. You will notice that small regions of the liver become completely perfused together when the solution is injected; these are the liver lobules.
3. Use clean scalpels and scissors to cut up the liver into chunks about 3 mm on a side—in other words, mince it. Keep it on ice the whole time.
4. Weight a sample of liver, about 5 grams, to the nearest 10 milligrams. Keep it cold and work as quickly as you can. (Remember to tare the weighing boat first.) Record this weight.
5. Place this weighed amount of liver into a homogenizer and homogenize it thoroughly, keeping it cold the whole time. If you are using a glass homogenizer (such as a Dounce or a Ten Broeck homogenizer), remember that the main grinding occurs as you work the pestle in and out; don't try to press it against the bottom of the tube. It may be most efficient to homogenize only a little bit at a time, adding more liver and more sucrose solution as you go along. Do not use more than 40 mL total sucrose solution or it will not fit in one tube.
6. Measure the amount of homogenate you have produced accurately to the nearest 1.0 mL by pouring it into a clean, cold graduated cylinder (and pouring it right back again into the homogenizer on ice!). Record the volume.

II. Low-speed Spin

This spin will sediment the nuclear fraction and leave everything else in the supernatant.

1. Pour your homogenate into a centrifuge tube and secure a second tube for a balance. Load the balance with water until the combined weight of tube, cap, and water equals that of the sample tube and cap to within 0.1 g. This is most easily checked with an old double-pan balance, using plastic beakers for tube holders.
2. Spin the sample opposite its balance tube for 5 minutes at 500 $\times$ g and decant the supernatant into a clean centrifuge tube. Discard the pellet, which should consist only of unhomogenized cells and connective tissue debris.

3. Spin the sample opposite a freshly weighed balance tube for 10 minutes at 1,000 × g. This produces three fractions:
   a) pellicle—partial cell membrane fraction
   b) supernatant—cytoplasmic fraction
   c) pellet—nuclear fraction

   Remove the pellicle carefully, using a Pasteur pipette, and discard it. Working quickly, decant the supernatant into a clean centrifuge tube and place the old tube in an inverted position for a few minutes to drain the pellet. If the pellet shows signs of slipping, place the tube upright immediately.
4. Resuspend the nuclear pellet in a minimum, measured amount of sucrose solution, preferably about 5 mL. Record the amount used, place the resuspended pellet in a storage tube, and label it as the nuclear fraction.

III. Second Spin

1. Take the cytoplasmic fraction (supernatant) from the previous spin and spin it opposite a freshly weighed balance tube for 30 minutes at 10,000 × g. This produces three fractions:
   a) pellicle—remaining cell membrane fraction
   b) supernatant—microsomal fraction
   c) pellet—mitochondrial fraction

Remove the pellicle carefully, using a Pasteur pipette, and discard it. Working quickly, decant the supernatant into a clean centrifuge tube and drain the pellet as above. Resuspend the pellet as above, in a recorded volume of sucrose solution.

This will probably take up an average lab period, and you will need to do the assays on these fractions in another period. Store all your fractions in the freezer for later use, being sure they are adequately labeled.

IV. Assays

A. DNA assay

The diphenylamine assay is a test for DNA. It works by hydrolyzing the deoxyribose with sulfuric and acetic acids, converting the sugar to an aldehyde that can react with the diphenylamine to produce a blue compound. The blue product absorbs light strongly at 600 nm, and the light absorption is proportional to the DNA concentration.

Procedure:

1. Precipitate proteins in each sample by adding an equal volume of warm 10% trichloroacetic acid (TCA). (*Note:* this produces very painful burns if it gets on your skin.) Mix the contents of your tube and place it in a hot (90°) water bath for 15 minutes. Shake it occasionally.
2. Centrifuge the tubes at 1,000 × g for 2 minutes and decant the supernatant. This supernatant can be used for the RNA and the DNA assays. Label it "crude nucleic acid extract" and proceed.
3. Diphenylamine assays require 4 mL of diphenylamine reagent (containing concentracted sulfuric and acetic acids—use caution!) for 2 mL of sample. The procedure:
   a) Pipette 2 mL of sample into a test tube
   b) Pipette 4 mL of diphenylamine reagent into the tube
   c) Cover the tube with Parafilm and invert to mix
   d) Remove the Parafilm and put the tubes in a boiling water bath for 10 minutes
   e) Cool the tubes quickly in an ice water bath

Carry out this procedure on your cell fractions and also on known samples of DNA to make a standard curve. (*Practical note:* the color developed in this assay is stable for several hours. Be sure to quench the tubes in ice water promptly after the 10 minutes are up, however.)

Suggested standard curve dilutions:

Final DNA concentration	Stock DNA solution (400 μ/mL)	5% TCA to dilute stock solution
0 (blank)	0 mL	2.0 mL
100 μg/mL	0.5 mL	1.5 mL
200 μg/mL	1.0 mL	1.0 mL
300 μg/mL	1.5 mL	0.5 mL
400 μg/mL	2.0 mL	0 mL

After all tubes have been boiled 10 minutes and then chilled, determine the absorbance at 600 nm for each tube, using the first tube in the table above your blank. If any of the cell fractions show unusual turbidity, centrifuge them before attempting to determine the absorbance at 600 nm.

B. RNA assay

This assay depends on the fact that hot HCl will convert ribose into furfural, which reacts with orcinol to form a green compound. Unfortunately, there is a weak reaction with DNA as well, so that about 10% as much color is developed with a mg of DNA as with one of RNA. If the DNA concentration in the extract is known, however, from the diphenylamine assay, the RNA concentrations can be corrected to allow for this.

1. Use the same crude nucleic acid extract as in step IV.A.2, above.
2. Set up the assay as follows: each tube should receive 3.0 mL of sample and 3.0 mL of orcinol reagent (caution: this reagent contains concentrated HCl) and should then be covered and mixed well. The tubes should be heated in a boiling water bath for 20 minutes (remove the Parafilm first). After the 20 minutes, the tubes should be quenched in ice water and the absorbance measured at 660 nm.
3. The standard curve should be constructed as for DNA, using a 400 μg/mL stock of RNA. Tubes will need 3.0 mL total sample, and 5% TCA is used as before to bring the final volume up to this. Suggested concentrations of RNA for the standard curve are: 0 μg/mL (blank), 200 μg/mL, 300 μg/mL, and 400 μg/mL. Fill in the table (top, right), using the rule that $V_1 \times C_1 = V_2 \times C_2$, where V_1 is the amount of stock solution taken and C_1 is its concentration, and V_2 is the final volume of the solution and C_2 is its concentration.

Final RNA concentration	Stock RNA solution (400 μg/mL)	5% TCA to dilute to 3.0 mL

C. Protein assay

Protein may be assayed using the biuret reagent; this is a copper and alkali complex that will bind to the peptide bonds in protein and cause an increase in light absorbance at 540 nm.

1. Using a fresh sample (not the nucleic acid prep, which has TCA in it) place 1.0 mL of each sample in a clean test tube and to each tube add 4.0 mL of biuret reagent. Mix well and allow to incubate at 37° for 20 minutes. Determine absorbance at 540 nm using as blank a tube with water and biuret reagent.
2. For a standard curve, use a stock of albumin of 10 mg/mL concentration. The dilutions should produce concentrations ranging from 1 to about 8 mg/mL.
   Outline below the dilutions for the standard curve.

V. Conclusions

You now have data on the concentration of protein, DNA, and RNA in your various fractions; recalling that you have volume measurements on each fraction and an original weight for your homogenized piece of liver, you can calculate the concentrations of these three molecules in the liver itself. Do this and record your results. You can find in most textbooks a table giving the composition of a "typical cell"—often a liver cell, in fact—and though this is usually given as a percentage of dry weight, it is simple enough to find data on the water content of cells so that you can compare these numbers with your data, which are based on wet weight. How close were you?

The Dance of the Chromosomes

You have probably come to realize that the cell is a complex thing with an elaborate structure; you have also, no doubt, heard about the arithmetical feat of cells, all of which "multiply by dividing." That musty old witticism tosses off an amazing process as if were just like pinching a lump of bread dough in two. But the structure of a cell is at least as complex as that of, say, a car—an you would hardly take it calmly if you saw a car produce a copy of itself! Even if we ignore the problem of how the individual parts are duplicated, there is still the question of how the intertwined, involved structures separate neatly and evenly. Yet the process goes on fairly rapidly and, in some cells, is repeated daily. How can this be?

We have only partial answers. Certainly the cytoskeleton is involved, and is restructured in the process. It seems that mitochondria can enlarge and pinch off portions of themselves like dividing bacteria and can glide along cytoskeletal tracks to distribute themselves evenly. Some organelles disappear. But one process, the separation of the chromosomes, is easily observed and much better understood, at least in some respects.

With DNA, the problem is packing. Each cell has a minimum of a meter of DNA (about 3 feet), and twice that before cell division. Of course, it is not in one piece, but there's good reason to think that it is not in very many pieces. Imagine having several dozen mile-long pieces of different-colored thread crammed into a tin can—and having to sort out all the colors and move half into a second can without using more than two cans worth of space!

The answer is packing—winding the DNA around histones and stacking up the histones until all the DNA is reeled into a manageable block. The block is what we call a *chromosome*, and the reeling is called *condensation*. (The histones are alkaline proteins that bind to the acid DNA.) Your cells have chromosomes all the time, of course, but their coiling is much looser and they appear as a smeary tangle of material called, *chromatin*, until they condense in preparation for cell division. Once condensed, the

chromosomes carry out a series of maneuvers designed to get the right number into each cell. Finally, in the two new ("daughter") cells, the chromosomes unreel again.

To see this process in a living cell is tricky. It is fast as cellular processes go, but still takes an hour or so, and special microscopes are needed to see detail in the live, unstained cell. But in any tissue sample used for studying dividing cells, there will be many cells at different stages of the process, so that, with a guide to the identification of the stages, you can "watch mitosis" in a bit of fixed and stained tissue.

Onion sets put down fine roots quite rapidly, and their root tips are good places to look for dividing cells. Other plants can be used, but thin, fast-growing roots are best.

The basic procedure for a simple stain:

1. Snip off the root from an onion set (take the whole root so nobody cuts off the stump later, thinking it is a tip), wash off any dirt, and put it in a drop of aceto-orcein on a slide—this is a good fixative and stain for chromosomes. Cut off all but the tip (3 mm).

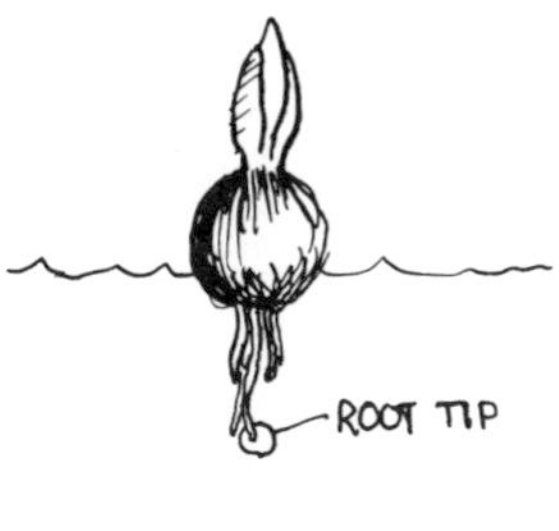

2. Add a coverslip, then squash the root tip by mashing a pencil eraser into the coverslip right over it. Mash firmly, or, if you can do it, flick the pencil against the coverslip with a dart-throwing motion.

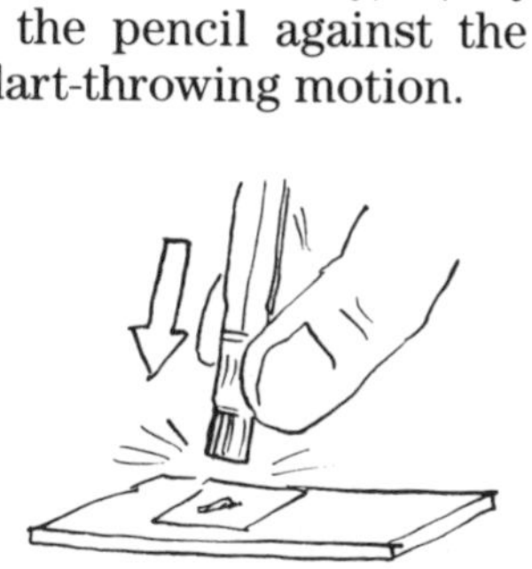

3. Examine briefly under low power of your microscope and determine which of the results below you have. (There isn't a "too much" category; plant cells are too tough to disrupt this way.)

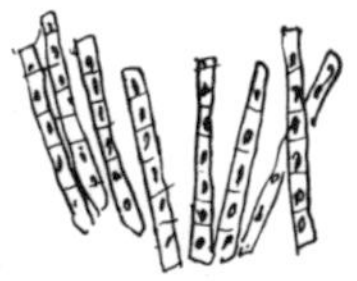

Your instructor may also want you to do a specific stain for DNA, the Feulgen reaction. This uses acid to reduce DNA to an aldehyde, and an aldehyde reagent (Schiff's reagent) to color the aldehyde reddish purple. Other substances in cells don't react, so you will be looking at the DNA alone. (Beware! Schiff's reagent is colorless itself, but will color many fabrics—like your clothes!)

The general procedure for a Feulgen stain:

1. Fix the root tips in ethanol:acetic acid 3:1—15 minutes to overnight, if needed.
2. Transfer tips to 5 *N* HCl, allow to sit 15 minutes.
3. Rinse acid from tips and cover them with Schiff's reagent until they turn deep purplish pink.
4. Wash off the Schiff's reagent with water and squash the tips as above, using 45% acetic acid in place of aceto-orcein.

With either staining method, if you can't get good squashes, try mincing the tip with a razor blade before squashing it. Be careful not to let the tip dry out at any time.

Use a higher magnification to identify "mitotic fiqures"—that is, cells in mitosis. Then identify each stage of mitosis. This is difficult because diagrams don't look like real dividing cells, and vice versa. Below is a chart showing both diagrams and more realistic drawings.

MITOSIS

Phase Name	Events	Diagram of One Chromosome	Diagram of Cell	Real Appearance
(Interphase	Cell growth, DNA replication		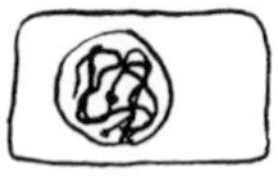	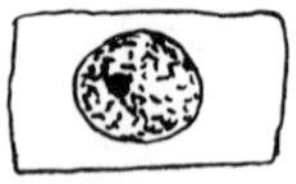)
Prophase	Beginning of condensation Towards its end, the nuclear membrane breaks down	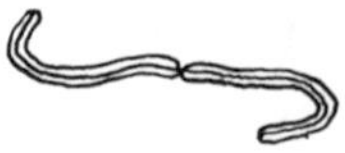 Chromosome has replicated already but 2 copies are joined . . .		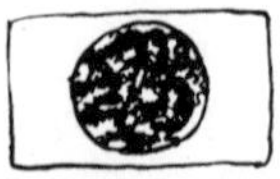 Filaments appear within the nucleus
Metaphase	Chromosomes arranged at central plane of cell, attached to spindle fibers	 Can now see 2 copies or chromatids but still joined at the centromere	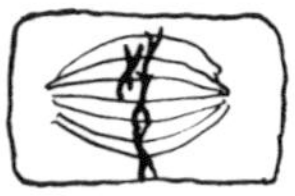	 Spindle may be difficult to see
Anaphase	Chromosomes begin to move apart—that is, the duplicated chromosomes split and the copies separate. They continue to separate until they reach opposite sides of the cells	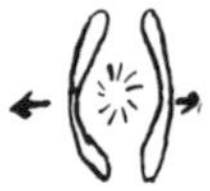 Pop! goes the centromere		 Mass of chromosomes beginning to separate
Telophase	Chromosomes de-condense, nuclear membranes re-form	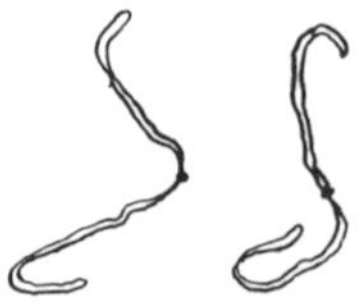 Each cell has one of these now. Go back to interphase		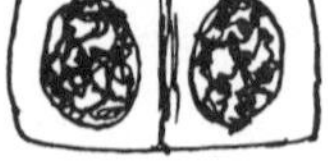

At the time of telophase, the cytoplasm also begins to separate, as shown above.

At first, you will find it easiest to identify anaphase, with its double cluster of chromosomes. Metaphase isn't as neat as you expect; the ends of chromosomes stick out all over. As the chromosomes condense and decondense, you will see a variable amount of ropiness in the chromosomes—at each borderline between phases, it becomes hard to assign a cell to one or the other, hence such forms as "prometaphase." The process is really a continuous one; the names of the phases are there as descriptive conveniences, not definitions.

You may have the opportunity to look at prepared slides of animal cells in mitosis, such as the cells of a section of a whitefish embryo (blastula stage), which divide rapidly. These will be essentially similar to the onion cells, with these exceptions:

The spindle fibers will be almost overshadowed by *asters*, one at each pole:

At metaphase you will may see cells from many angles, and sometimes they will have a wreath of chromosomes:

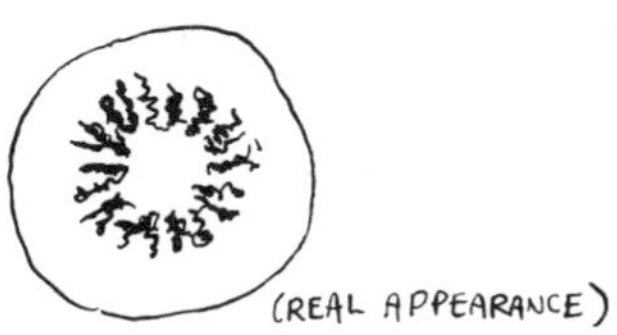

Telophase will be accompanied by cell division, as in plants; but instead of building a partition between cells, animal cells pinch in their outer membranes:

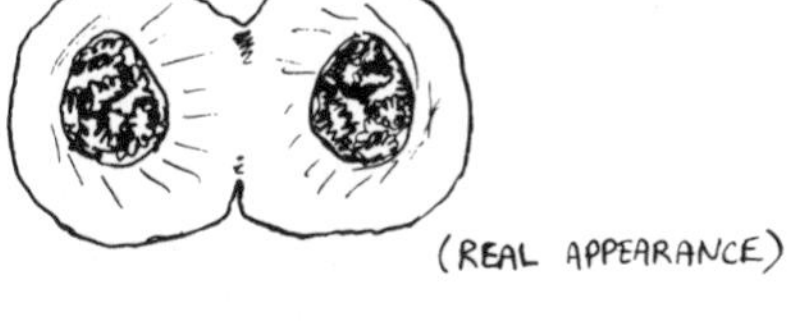

Aside from these differences, plant and animal cells divide similarly. From each, find a good example of each phase and draw it. You will notice that the chromosomes are not discernible as individuals but are packed in a mass (unlike diagrams). To see chromosomes as individual structures, for example in genetic studies, it is necessary to "trap" cells in metaphase (using a mitosis-arresting drug like colchicine) and cause them to swell up by soaking them in water; then, when the swollen cells are gently ruptured on a slide, they spill out their chromosomes in a circle large enough to avoid overlapping. You may have a demonstration of a slide prepared in this way, or of a *karyotype* (a chromosome picture) made from one.

Now consider: the misplacing of one such chromosome has terrible effects on the whole organism (most trisomic embryos don't make it), and it would seem that it probably isn't good for single cells, either, although cancer cells do tend to have abnormal numbers of chromosomes. The dance of the chromosomes—mitosis—is the sorting-out process that ensures the regular allotment of genes.

Of course, it wouldn't be fair not to tell you that there are certain insects (lowly creatures with no respect for Science!) that happily throw away most of their chromosomes during mitosis in every cell except their sex cells. *That* is a subject best left alone!

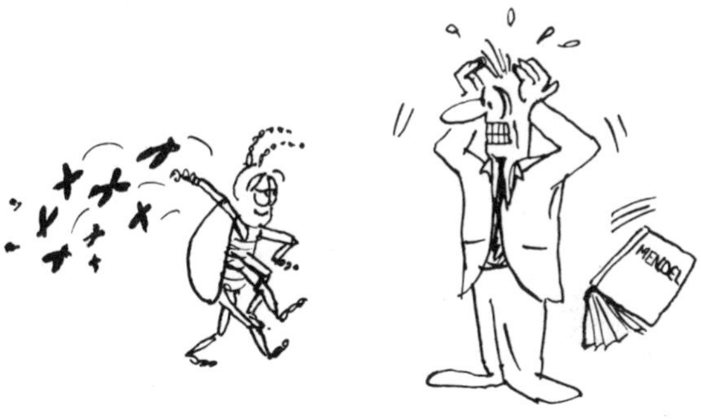

Shuffling Genes

TICKETS
THEY ALL YOURS?
HOPKINS

Mitosis, as you have seen, is a very orderly, regular process. It produces two identical cells in a delightfully predictable way. But it *is* a bit monotonous. If this were the only way in which cells could divide, then all organisms in a species would be carbon copies of their parents (except for mutation).

In spite of the way some families look, this doesn't happen. Metazoans have developed a way to ensure variations—sex. Getting two individuals to make genetic contributions to each new individual makes a more pleasingly varied population, and it seems that this variation is valuable; physical disasters or epidemics are less likely to destroy all of a diverse population, just as rainy weather is harder on a paint store than on a general hardware store.

Now sex is a great solution, but it isn't as easy as it seems. How can two individuals make genetic contributions to one new one? Certainly not in the obvious way, by pooling genes. Your genes reside in your 46 chromosomes. Combine them with someone else's, and the result is a double set of genes and 92 chromosomes. Although plants sometimes do this (and produce giant blossoms when they do, to the delight of florists), animals can't get away with it. Even one extra chromosome results in fetal death or severe birth defects. And in any case, with the number of chromosomes doubling in each generation, sooner or later there would be a very loud POP! Before sex, there has to be a reduction in chromosome number.

We do this by *starting* with a double set and, before sex, reducing it to a single set. The double set consists of 23 pairs of *homologous chromosomes* (or *homologues*), one of each pair from one parent, the other from the other parent. The genes are arranged along both homologues in the same order, though they are different in the type of information they carry—blue vs. brown eyes, and so forth (as shown). It is all rather like a formal dinner, in which the appetizer always comes first (though it may be shrimp or paté or oysters) and the soup next (though it may be onion or chowder or bisque), and so forth. "Gene" is not often used in this context, but instead *locus* (appetizer vs. soup, etc.) and *allele* (shrimp vs. paté, etc.). So, on our diagram, the third locus down is the "nose-length" locus, and we see two alleles, long and short (more may exist).

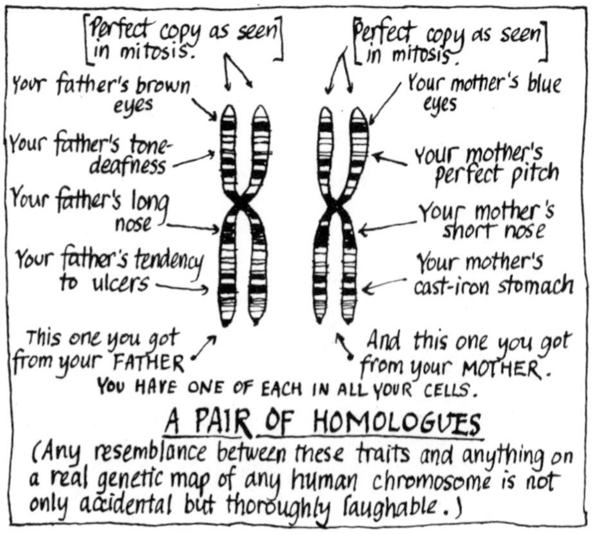
Perfect copy as seen in mitosis.
Perfect copy as seen in mitosis.
Your father's brown eyes
Your mother's blue eyes
Your father's tone-deafness
Your mother's perfect pitch
Your father's long nose
Your mother's short nose
Your father's tendency to ulcers
Your mother's cast-iron stomach
This one you got from your FATHER
And this one you got from your MOTHER.
YOU HAVE ONE OF EACH IN ALL YOUR CELLS.

A PAIR OF HOMOLOGUES
(Any resemblance between these traits and anything on a real genetic map of any human chromosome is not only accidental but thoroughly laughable.)

With this in mind, consider another problem: if you must choose between homologues and bequeath only one to your offspring, then genetic variation will be limited by the number of chromosomes, each acting as an indivisible packet of information. This is still a lot of variation (figure out what 2^{23} is!), but we would like more. The "more" comes from *recombination*, a process in which homologues pair up, locus by locus, and swap bits here and there—so that, in our example, the brown eyes could end up on the same chromosome as the perfect pitch.

TRADE YA A BLUE-EYES FOR A BROWN-EYES ?

(And who knows—a producer could be looking for a brown-eyed singer!) Of course, if the homologues happened to have the same alleles at the exchaged loci, it wouldn't make any difference; but that is very unlikely.

Meiosis in the Cricket

To see meiosis one must look at a creature's gonads, since it only a way of producing gametes (sperm or eggs). Crickets are a good source, and males are best in any species, since they produce greater numbers of gametes. Pick out a male:

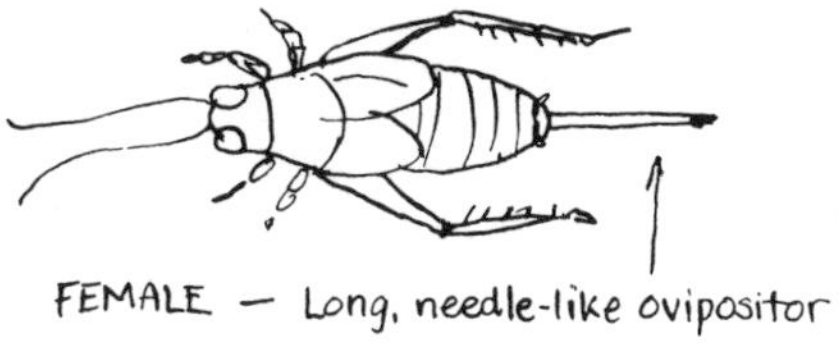

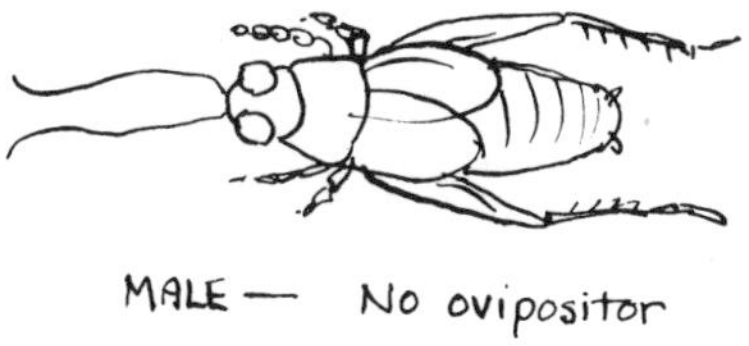

Put it in a jar with a bit of chloroform-soaked cotton, or put it in the refrigerator, to anesthetize it. Then, holding it under water in the dissecting pan, snip open the abdomen along the ventral side. The largest structures in it will be two white, soft spheres—the testes. Drop the testes into 45% acetic acid and mince them up, or tease them apart with dissecting needles. Transfer a few bits to a drop of aceto-orcein on a microscope slide, and after a few minutes add a fresh drop of stain; then cover the slide with a coverslip and squash the testis fragments with a pounding from an eraser. As in all squashes, try to avoid side-to-side movement.

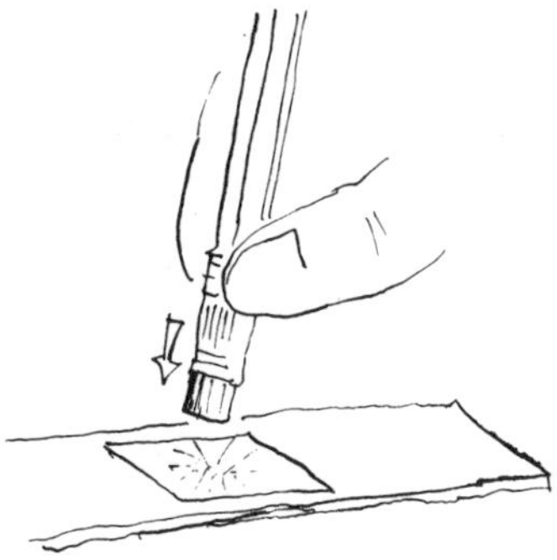

Examine the slide, first under low power to locate areas of uniform red patches (the nuclei) and then under high power to identify the stages. If your preparation seems understained, seal the edges of the coverslip and wait a few minutes. You can use Vaseline or nail polish for this.

You may have prepared slides of sectioned grasshopper testis; compare these with your squashes and pick out any stages you can't identify.

Meiosis, you will see, does all the things we mentioned as required for making sex cells; it *reduces* the number of chromosomes and allows for *recombination* between homologues. It goes in two stages, like a double mitosis. Here is a chart of it:

Meiosis

First, it is helpful to remember that nuclei of cells in meiosis I are longer than those of cells in meiosis II; this is an aid to identification.

MEIOSIS I

Phase Name	Events	Diagram of One Chromosome	Diagram of Nucleus	Real Appearance
[Interphase]	as in mitosis-DNA replication (but 2 copies not visible)			
Prophase Leptotene	has 5 stages: Beginning of condensation			
Zygotene	Homologues pair		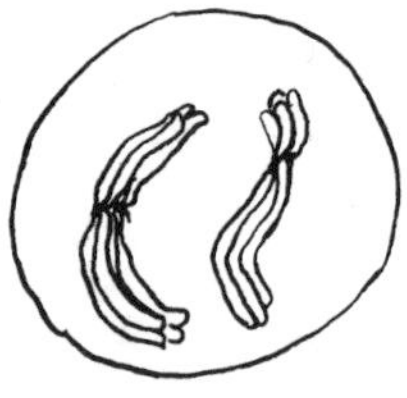	
Pachytene	Paired homologues thicken & shorten. Recombination occurs.		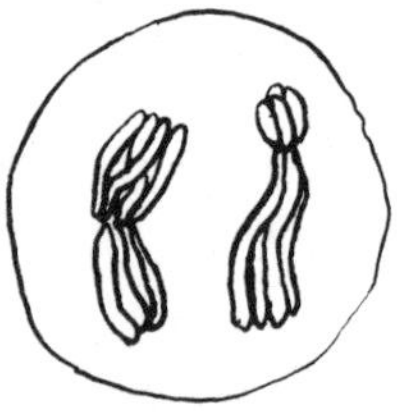	
Diplotene	Homologues begin to separate, but seem to stick at places where recombination occurred: these places are *chiasmata.* (Also earliest point at which 2 copies of each become visible.)	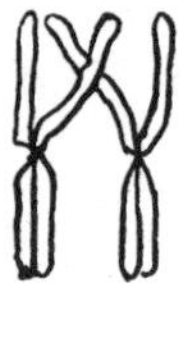		
Diakinesis	Homologues separate more; chiasmata move to ends, forming V & O shaped figures.	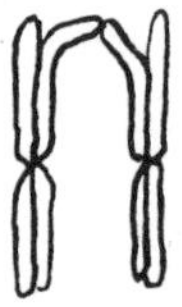	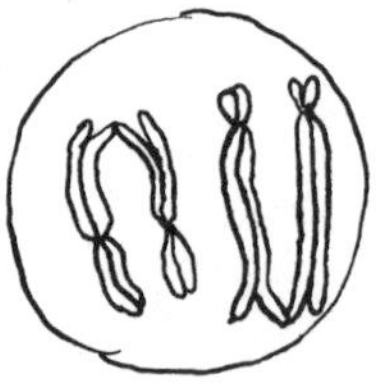	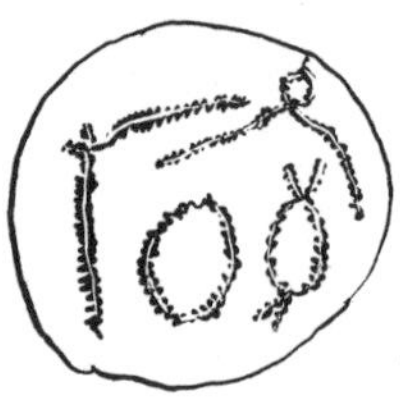

Phase Name	Events	Diagram of One Chromosome	Diagram of Nucleus	Real Appearance
Metaphase I	Homologues line up in middle of cell.	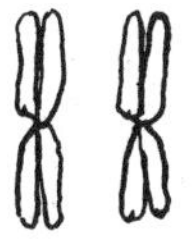	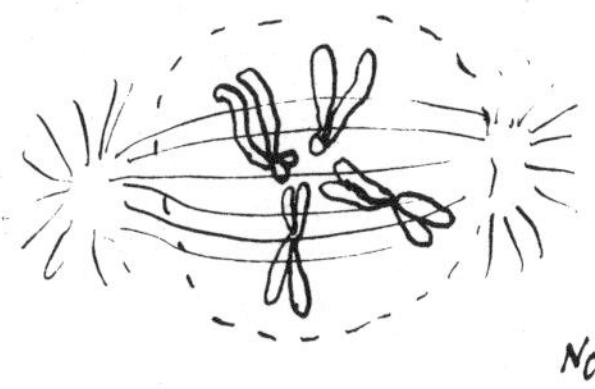	
Anaphase I	Homologues separate, not copies as in mitosis.	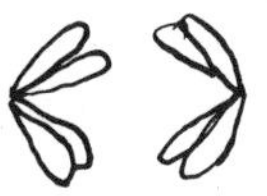	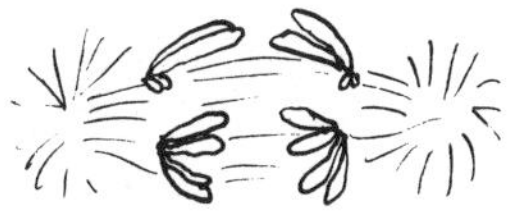	
Telophase I	Chromosomes de-condense, nuclear membrane re-forms.	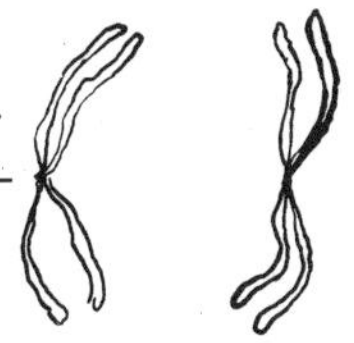		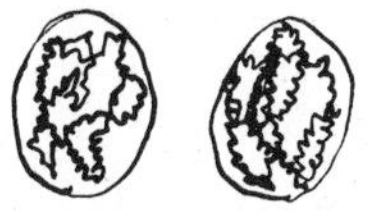

Now there are two cells, each with one of each homologous pair. Each homologue has already replicated in the previous interphase, but the copies are still attached at the centromere (notice the  shapes in anaphase & telophase above, compared to the  seen in these phases during mitosis).

MEIOSIS II

–now following one daughter cell from meiosis I

[Interphase–appearance is like that of telophase above, as chromosomes usually do not de-condense as much as in other interphase. *No DNA replication* in this phase. It's just a brief pause before . . .]

Phase Name	Events	Diagram of One Chromosome	Diagram of Nucleus	Real Appearance
Prophase II	chromosomes condense, no pairing.			
Metaphase II	Now chromosomes line up at center of cell . . .		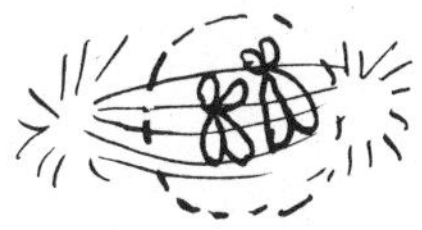	
Anaphase II	. . . and the copies separate at the centromere, as in mitosis . . .			
Telophase II	. . . forming two *haploid* cells.	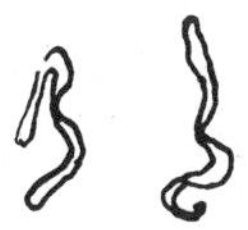		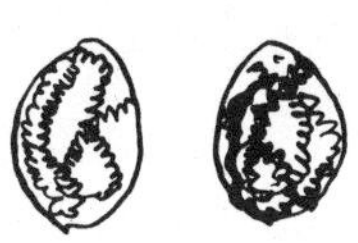

Questions:

1. Which of these steps provides for genetic variation in the sex cells? Which provides for reduction of chromosome number?

2. Why do you suppose the chromosomes are so extended in meiosis I compared to meiosis II?

Genes in Human Populations

We can study the genetics of peas, as Gregor Mendel did, or of fruit flies, as Thomas Hunt Morgan did, and learn quite a bit. Of course, you go through a lot of peas (or flies), and you may lose a few friends in the process—but it does work. People often ask why geneticists study such organisms as these—why not study people? That would be so much more useful. But there are a few problems with this; consider the classical methods of genetics. First you must have true-breeding animals (or plants), obtained by generations of brother-sister matings. Then you must set up test crosses between these different stocks, followed by a second generation of brother-sister matings or, frequently, a backcross—an offspring-parent mating. All of this, applied to humans, might make a steamy novel indeed, but it wouldn't sell in Peoria! Not all of our matings are legal, perhaps, but we at least like to choose them ourselves.

We're stuck, then, with a messy, uncontrolled system if we study human genetics. We do detective work rather than experiments. Fortunately, we have learned the basic rules of the game from those supposedly irrelevant studies on animals and plants, and we can reason deductively to find out what is happening in humans. We have three main lines of evidence:

1. Pedigrees: More politely, "family histories," these are written up on standard forms with circles for females and squares for males, each one colored in according to whether or not the individual had the trait in question. Because so many people in a pedigree are dead or otherwise unavailable, the traits must usually be something fairly obvious, which would be remembered clearly or which would show in old pictures. Pedigrees based on less obvious traits are much shorter and more fragmentary.
2. Biochemical analysis: There are many enzymes and other proteins that have genetic variation among humans. Sometimes these are related to disease; for instance, hemoglobin S, which is found in people with sickle-cell anemia. Or sometimes they are important for other reasons, as are the proteins called HLA, which play a

major role in graft acceptance or rejection. But often the value of a genetic variant is purely as a *marker*—a signal that the person got that gene, and presumably others near it, from one particular parent. The significance of that particular gene itself may be nothing.

3. Karyology: White blood cells of one type (lymphocytes) may be taken from a small blood sample and stimulated to divide by means of various drugs; then they may be fixed and stained, and inspected for mitotic cells. Those in metaphase (see the procedure in the lab on mitosis) may be photographed and the chromosomes cut out of the picture and assembled in a standard sequence, a *karyotype.* This can then be compared with those of other people, both normal and abnormal.

There are certain shapes and banding patterns on chromosomes that permit them to be used as markers, in a sense; and, of course, any abnormalities in a karyotype are of great genetic significance.

Human Traits

Here is a sampling of easily detectable human traits which are genetically based. In some cases one trait is clearly dominant over its alternative, just as in Mendel's peas. In other cases the alternatives can be codominant—both expressed, as in the blood group ABO. And in some cases the picture is more complicated still, as a particular apparent trait (phenotype) is the result of many genes' action (polygenic) and may behave quite erratically.

For each characteristic, record your own phenotype. When you know it, make a note also of your parents' phenotypes and those of any other close relatives. If you have the opportunity to do so, continue this on your next visit home (your instructor will provide you with taste papers) and construct a pedigree.

Anatomical Traits

Long palmar muscle: This is one of the most variable muscles in the body; about 10% of the population lacks it. To find out whether you have it, clench your fist tightly and flex your wrist. Count the tendons that you can feel just above the wrist. Most people have three, but a dominant gene (it seems) is responsible for the absence of the middle tendon, that of the *palmaris longus* muscle.

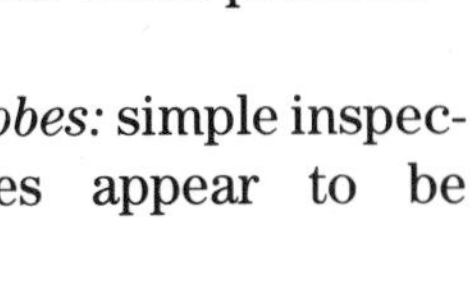

Darwin's ear point: An extra point in the cartilage of the outer ear is another rare but dominant trait. It is quite prominent when present.

Attached or unattached earlobes: simple inspection again. Attached lobes appear to be recessive.

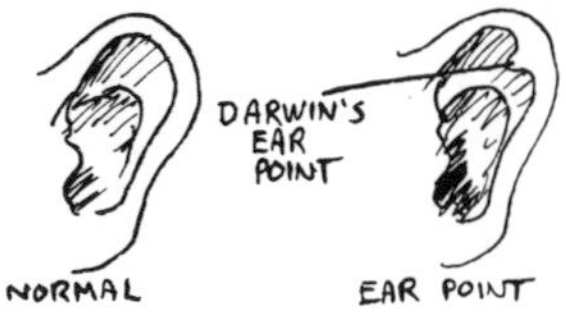

Second finger shorter than fourth: This rather distinct trait is influenced by sex. Is it more common in the males of your class, or the females?

Tongue rolling: This is a dominant trait—the ability to extend the tongue and roll it into a U-shape as shown.

Hitchhiker's thumb: The ability to hyperextend the thumb may be due to a recessive gene.

Pigmentation

Albinism: Melanin, a brown-black pigment, is responsible for hair, skin, and eye color, and everyone has it (distributed differently) except true albinos, who are homozygous for a recessive gene.

Eye color: Blue- or gray-eyed people appear to be recessive for a gene that determines whether melanin will be deposited in the front part of the iris. All other colors appear dominant (and differ only in amount and distribution of pigment).

Hair color: This is complex, since it is affected by the microscopic structure of the hair as well as the amount and distribution of melanin. One hypothesis considers two genetic loci, one for dark vs. light hair (with light as recessive) and another for red vs. nonred hair (with red as recessive and also concealed—except for highlights—by dark). How does this hypothesis fit with your pedigree?

Chemical Traits

Taste shows many variations, because those sensitive chemical sensors, our taste buds, are genetically programmed.

Phenyl Thiocarbamide (PTC): This is a harmless substance that some people cannot taste, while others perceive it as bitter or as some other disagreeable flavor. Tasting is dominant over nontasting. Suck briefly on a bit of clean filter paper, to get used to its taste; then suck on a bit of PTC-soaked paper. Can you taste it? If you are a taster, get together with other tasters and discuss your food likes and dislikes; compare them to those of nontasters. There are several compounds that can be tasted by some people and not by others; tasting is usually dominant. Try any other such compounds that are available and record your results. In discussion with the class, can you draw any general conclusions about the relation of these tasting abilities to food preferences?

Excretion of thioesters: This is a striking metabolic trait. Some people, after eating asparagus, excrete S-methyl thioarcrylate and other thioesters in the urine. These compounds are toxic in high concentration, and very malodorous. You undoubtedly know it if you are an excretor.

Blood type: Blood group substances show various patterns of inheritance; for example, in the ABO system A and B are codominant to each other, and O is recessive to both. If you have not done blood typing yet, follow the directions given in "The Cell Surface." Do a pedigree on the ABO and Rh systems; most of your family probably know their blood type or have it on a card if they have given blood. The usual way of giving it, say, as "A negative," means ABO type A, Rh type negative.

"Dictator"

Now take a few minutes to play a strange little game that we will call "Dictator." Select some trait that everyone has been able to score, and that has some dominant and recessive alleles; say, eye color. Now take two counters—they may be black and red checkers—according to your own genetic makeup. If your eyes are brown and so are your parents', take 2 black checkers; if yours are brown but you have a blue-eyed parent, take one black and one red, and so forth. These are your alleles.

Now, select a dictator. Perhaps your instructor is a natural choice. This person, we pretend, is determined to eliminate blue-eyed people. He or she can "execute" anyone desired and compel any two players to marry and produce offspring. But the dictator can't see your two checkers; you only make public your phenotype. Each time a mating is ordered, you and the other player place both sets of checkers in a bag, shake, and each draw out a pair. These are the alleles of the offspring. (To increase accuracy, you could just randomly combine one checker from each person.) In this new generation, "you" will die (for simplicity's sake) and assume the phenotype of your "offspring." (Only two offspring per pair allowed!)

Now let the dictator do his or her worst, during as many generations as you have time for. Play this in different traits, and then discuss.

1. Is there a difference between trying to eliminate a dominant trait and trying to eliminate a recessive one? What is it, and why?
2. Suppose your dictator tries to *increase* a rare trait (rather than eliminating it). How does this change things?
3. The serious ideas behind this game have been around for 200 years and are reviving today. What do you know about modern attempts?

Getting It Together: Fertilization and Development

All of the complexity that is you began as two cells that fused into one. That kind of statement gets made so often in biology texts that you probably read it without a thought. Try saying, "This house began as a single brick," or, better, "This computer began as a single blob of solder," instead. How in the world can a single, simple subunit build itself up into a billion-subunit, integrated complex? It is the oldest and best example of pulling yourself up by your own bootstraps. It has no business happening; it shouldn't work. But it is constantly happening—all those cells, as if ordered around by some invisible, perfectionist supervisor, move, change, move again. They form patterns, then dissolve them; gradually the shifting structures become familiar: a new organism is forming. The world has never seen organization like the organization that accomplishes *that!*

So the question, as usual, is, how does it work? If you are expecting a simple answer, don't hold your breath. It seems clear that there are a number of very different stages in the process, and that the early ones lay the groundwork for the later ones. The first few divisions of the fertilized egg establish a structure in which the different cells gradually become oriented, and it is this orientation that determines the part they play in the rest of development. From there on, different organs develop more or less independently, so the picture becomes more complicated. Let's go back, then, to the early stages, and concentrate on them.

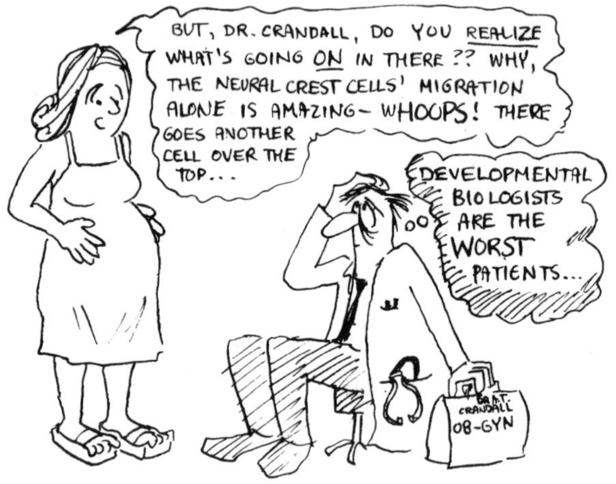

It all starts with a bang—the tiny collision between egg and sperm. Cells with a half-complement of chromosomes, they are the products of meiosis and *gametogenesis*, a process in which future egg cells are stuffed to the bursting point with rich yolk and messenger RNA, while future sperm are stripped and slimmed down to essentials and acquire a powerful tail for swimming.

The collision between egg and sperm is not a simple, mechanical event, however. There are, it seems, signals given and received on both sides, and a whole series of active responses made—a sort of chemical courtship.

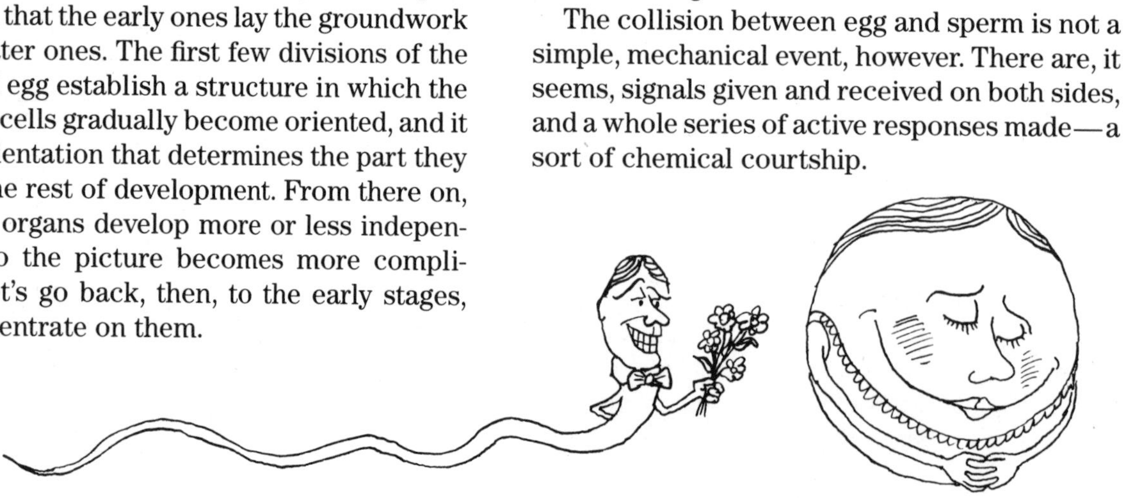

Fertilization in the Sea Urchin

In the nineteenth century, German embryologists started going to the Riviera—strictly for scientific purposes, for course, since they wanted to use sea urchins for a study of fertilization, and there are plenty of sea urchins there. Fertilization is external in echinoderms, and easy to control.

You can tell the sex of a sea urchin only by the color of the sex cells it releases, so just pick one at random (the spines are *not* dangerous) and turn it over. The five white teeth (which will probably open and close slowly) are surrounded by a ring of nonspiny skin, the *oral membrane.* Fill a 1-mL syringe with the KCl solution and insert the needle into the oral membrane, but sloping slightly away from the mouth. Inject the whole 1 mL, slowly. If you have

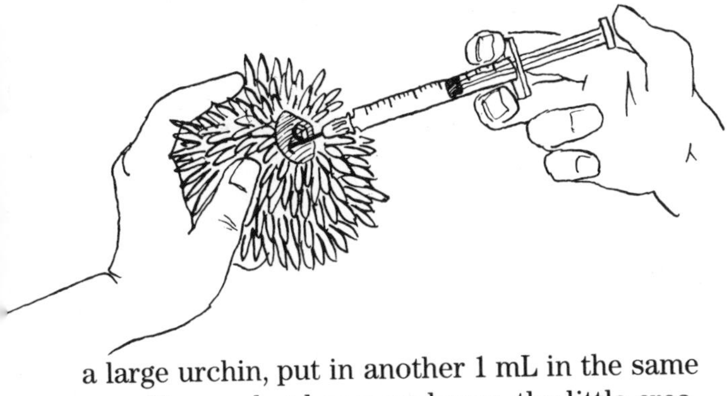

a large urchin, put in another 1 mL in the same way. If your shot has gone home, the little creature will begin to wave its spines around quite vigorously. This is a prelude to a great gush of gametes, so put the urchin down, still mouth side up, on a clean glass petri plate. Be sure the plate is dry. Meanwhile, get a 100-mL beaker and put seawater in it, about 25 mL.

By this time, your urchin should be shedding gametes. Look at what has collected in the plate; it if looks granular and is colored—red or gold, usually—it is eggs. Put the urchin on top of the beaker of seawater and let the egs fall into the water. If, on the other hand, what is in the plate is thick and milky white, it is sperm; leave the urchin where it is and don't let the sperm touch seawater. If nothing happens, try another urchin—some are just not fertile.

In a few minutes, when they seem to have stopped shedding, remove the urchins and prepare the gametes.

Eggs: Pour off the water the eggs are in, leaving the eggs in the bottom; label this "egg water" and save it, and replace the liquid on the eggs with fresh, clean seawater. Gently agitate the beaker to suspend the eggs, and then let them settle out again. Pour off the water (discard it this time) and replace it with fresh again; do this two more times. Make sure that the eggs are cool; don't leave them under hot lights, for example.

Sperm: Sperm must be diluted about 100-fold in fresh seawater before use, but once this is done, they are good for only 20 minutes. The best procedure is to take 10 mL for seawater in a beaker and, just before you need them, add 1 drop of sperm. Keep this diluted sperm cool, and write on the beaker the time it was diluted.

You will also need some depression slides, or else a syringe full of petroleum jelly with which to make corner supports to keep the coverslip from crushing the eggs (which are large). In addition, have nearby a test tube with a little of the egg water, and a supply of dropping pipettes.

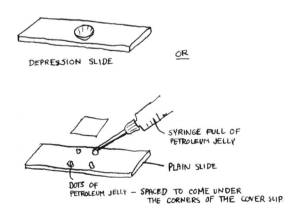

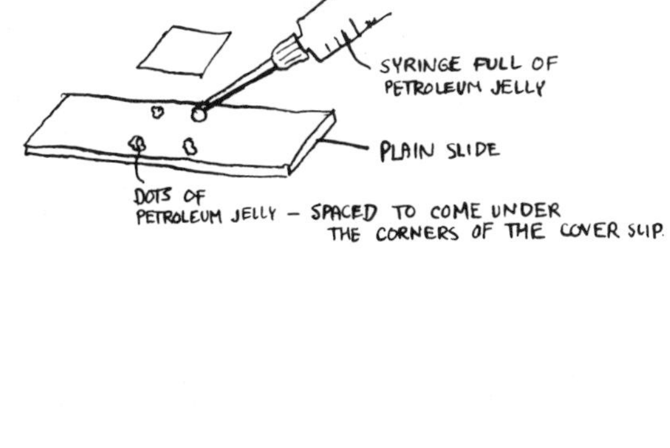

Fertilization

Take a drop of the egg suspension and dilute it with a milliliter or so of seawater; then inspect it under your microscope under low power. The eggs will be easy to identify. You should have a number of eggs on your slide—but not too many; a few dozen on the slide at most. If you need to adjust the number, add more egg suspension or more seawater to the slide.

Now, focus on several eggs, stop your condenser diaphragm down (less light and more contrast), and add a drop of diluted sperm to the edge of the coverslip. Immediately begin observing the eggs. As the sperm swim into your field of view, you should see their heads quite clearly, and, with good condenser adjustment, their tails also.

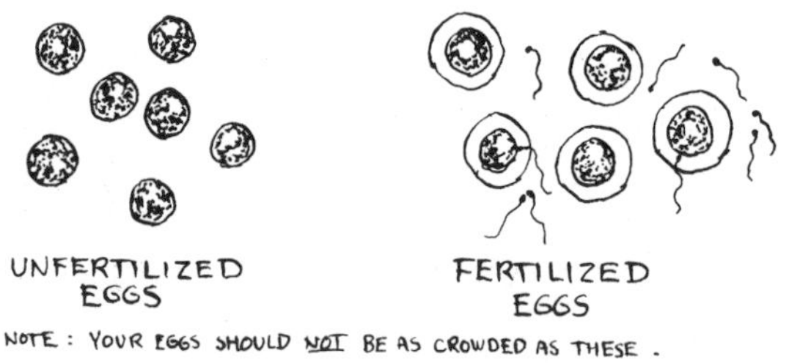

Are they swimming in a particular direction?

What happens when they contact the egg?

As you can see, the egg has an outer coat that is sticky for sperm. Often you will see captive sperm, still swimming, lashing their tails and spinning the egg around. In a minute or two after fertilization, a dramatic change takes place in the egg: a thick outer coat is elevated from the egg surface. This is the *fertilization membrane.*

What proportion of the eggs is fertilized?

What becomes of the extra sperm?

Set the eggs you have fertilized aside, placing them in a moist, cool place, like a slide box with a moist paper towel in it. After about an hour, take another look at it.

Sperm Behavior

Repeat the above fertilization, but put a drop of diluted *sperm* on the slide first; focus on them and note how they are moving. Is there any direction in which they are going? Now add a drop of *egg* suspension at the edge of the coverslip and immediately observe the sperm. How does their movement change? Remember that the microscope image is reversed.

Does this confirm any earlier observations?

You see that this implies a chemical attractant of some kind; if it traveled in air, we would call it a smell. But this particular "smell" is present in such small amounts that its identity is still a matter of controversy.

But here is something stranger: put a drop of sperm suspension (by this time you must have had to make up fresh suspensions) on a clean slide; to it add a drop of water from the tube you labeled "egg water"—or from any other seawater in which unfertilized eggs have been standing for a while. Now look at the sperm. What a mess! Can you tell what particular parts of the sperm are being stuck together?

It appears that there is something released by eggs that can clump sperm cells together in the same way that your blood-group antisera could clump red blood cells. But why? Is it a way of thinning out the sperm? Polyspermy—being fertilized by more than one sperm—is not good for an egg. Or is it an excess of the stuff that sticks the sperm to the egg surface? It could be both.

Development

An hour after fertilization, more or less, your eggs will start dividing, forming the first of the many cells of the new individual. But these divisions are not accompanied by growth; the egg was a big cell, and at first it is simply subdivided until the component cells are normal-sized. These divisions are called *cleavage*, and they begin the process or orienting the embryo, setting aside particular cells for particular functions. In another half an hour, the two cells will cleave into four, and so forth. If you can keep your slide in a cool, moist place and come back later to look at it, you may find some of the stages below. You may also have access to prepared slides showing the same stages of development in the starfish. If so, identify the stages in those slides.

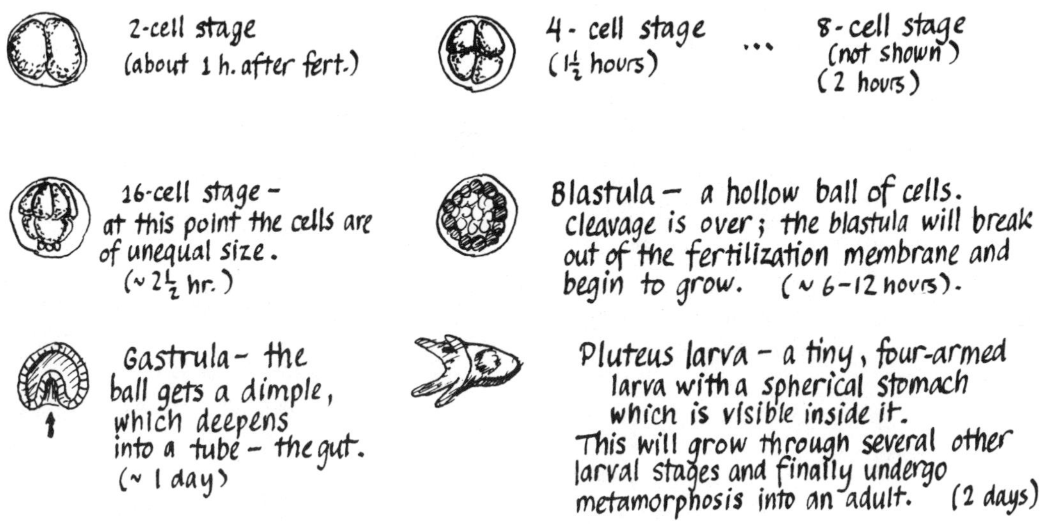
2-cell stage
(about 1 h. after fert.)
4-cell stage
(1½ hours)
...
8-cell stage
(not shown)
(2 hours)
16-cell stage - at this point the cells are of unequal size.
(~ 2½ hr.)
Blastula - a hollow ball of cells. Cleavage is over; the blastula will break out of the fertilization membrane and begin to grow. (~ 6-12 hours).
Gastrula - the ball gets a dimple, which deepens into a tube - the gut.
(~ 1 day)
Pluteus larva - a tiny, four-armed larva with a spherical stomach which is visible inside it.
This will grow through several other larval stages and finally undergo metamorphosis into an adult. (2 days)

Questions:

1. Why do you suppose you had to prepare a fresh sperm suspension every 20 minutes, but could use the same eggs for hours?
2. Given the sorts of observations you have made, could you devise any ways of finding out some characteristics of the clumping factor (which, by the way, is called *fertilizin*)? Look at the exercise "The Cell's Alchemists" for ideas.

Symbiosis

A good many animals make a living from other animals, not as predators but as symbionts—as the name implies, ones who live together. Sometimes symbiosis is beneficial to the symbiont and the host as well. For example, there are protozoa in the guts of termites; they are protected from the outside world and provided with lots of chewed-up wood by the termite. In return, they digest the wood, which termites *cannot* do, and leave some of the digestion products for the host. This sort of happy arrangement is called *mutualism.*

Sometimes symbionts are just harmless freeloaders who neither help nor harm their hosts, so far as we can tell. There are protozoa in the human gut that produce no ill effects, for instance. They evidently make a living sharing our meals with us. All such are termed commensals, for *commensalism* means "sharing meals."

But most symbionts are not so restrained; they chew their hosts' tissues, suck their blood, or simply eat so much of their food that they starve. When symbiosis is injurious to the host, it is called *parasitism* (from the Greek term for a social hanger-on). Most parasites must face the problem of what to do when the host is finally used up. The usual solution takes the form of a complex life cycle in which each stage is spent in different species of host: the parasite just keeps moving along, like the Mad Hatter's tea party.

Lest we become too proud, we should remember that at least one fellow mammal, the vampire bat, is a parasite, and some suggest that man, too, became one when he changed from hunter to herder!

Mutualism in the Termite

Place a termite on a slide in a drop of insect saline and, viewing it under a dissecting microscope against a dark background, place one dissecting needle on its head, and another on the tip of its abdomen and pull away the latter, as shown.

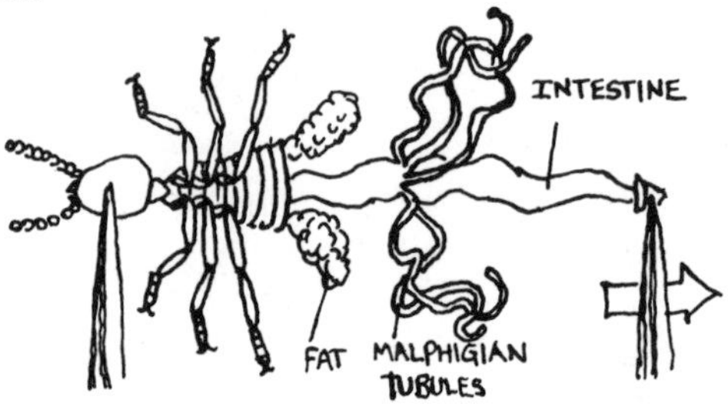

Discard the hard exoskeleton but save the intestine. The other soft parts will not need to be removed. Tease apart the intestine with your dissecting needles and cover it with a cover slip. Examine this slide under a low-power objective of your microscope to locate movement, then change to higher powers. Be sure your condenser is set up properly; use phase contrast if you have it.

You may see some or all of these protozoan symbionts:

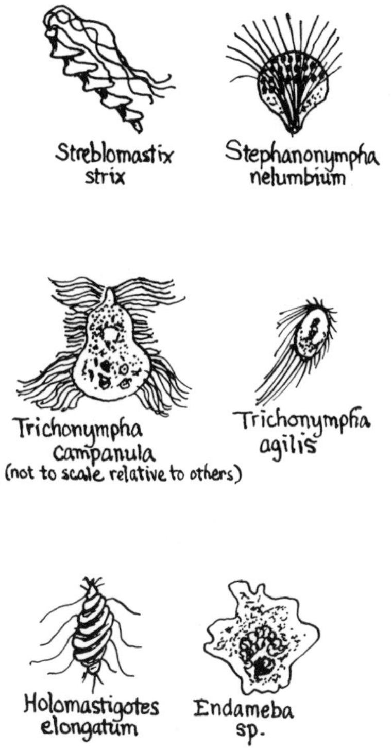

As in all intestines, you will also see legions of bacteria.

Sketch and try to identify the organisms you see. Note their relative sizes and the proportionate numbers of each type. Do you see anything that might be ingested wood bits in any individuals?

Questions:

1. Were the organisms you observed motile? Why do you suppose they have retained this ability?
2. Some treatments, such as heating to 36° C, will kill the protozoa without injuring the termite. What future would you predict for termites so treated? Could you rescue them?

Parasitism in the Frog

Get a pithed frog from your instructor and pin it in your dissecting pan, ventral side up. Work quickly as you dissect—you should be familar with the vertebrate body plan by now—and remember that the parasites you are seeking will begin to die as their host does unless they are removed to fresh saline soon.

Keep the frog covered with moist paper towels when you are not working on it.

Have on hand several dishes of Amphibian Ringer's solution and a stock of clean slides.

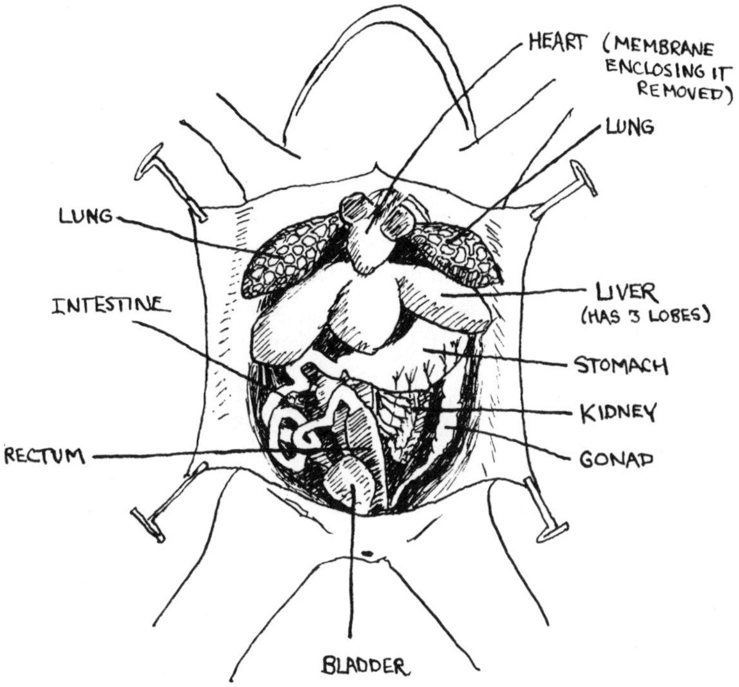

Lungs: Remove both lungs as soon as you have opened the frog. Place them in a single dish of saline and gently open them with scissors. Tease the the tissue apart with dissecting needles, observing it under a dissecting microscope. You may expose some red-brown moving organisms up to 8 mm long. These are lung flukes, Phylum Platyhelminthes, Class Trematoda. If you find one, put it on a slide in Ringer's and examine it under low power on your microscope. These parasites live in the lung and shed their eggs into the glottis, where they are swallowed, pass out with feces, and are eaten by snails. In the snails the eggs give rise to a larval fluke, which reproduces asexually; its progeny invade the rectum of a dragonfly nymph, and when it leaves the water as an adult, the flukes go along. When the dragonfly becomes a frog's dinner, the cycle has been completed!

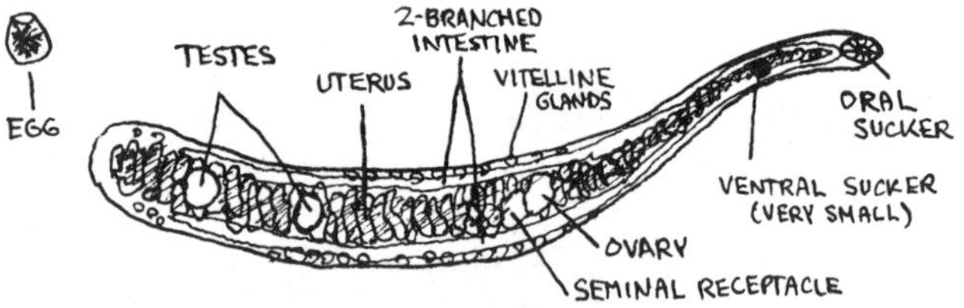

Haematoloechus medioplexus (lung fluke)

Bladder: Remove the bladder and dissect it as you did the lungs. The common bladder fluke is small, less than 3 mm long. If you find one, mount it on a slide and study it as you did the lung fluke. Its eggs resemble the lung fluke's, and are also shed into the water, but in this case the next host is a clam, and after that a variety of small mollusks, which are eaten by the frog. The young flukes travel along the gut until they reach the cloaca (the common digestive and urinary tract outlet), where they make a U-turn to home.

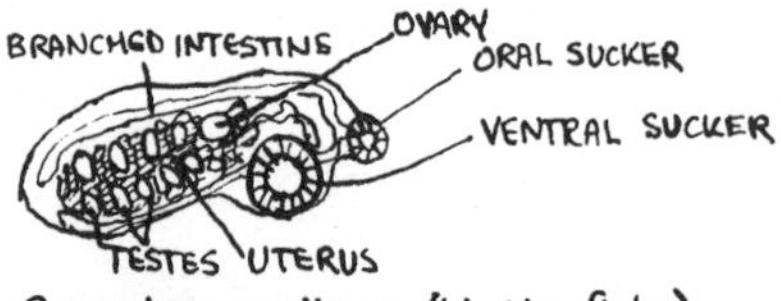

Gorgodera amplicava (bladder fluke)

Blood: Have a clean pipette ready, and several slides. Use your scissors to cut the blood vessel leading to the kidney, and pipette up some of the blood; make several slides of blood, diluting some with saline severalfold. In the undiluted blood, look for a small flagellate swimming among the blood cells. It has a relative, *Trypanosoma gambiense,* which causes sleeping sickness. This trypanosome, *T. rotatorum,* cycles between frogs and leeches, being transmitted by the bite of the leech. Also, you may see other protozoa that live inside the frog's red blood cells; use the diluted blood slides for this and adjust your condenser with care.

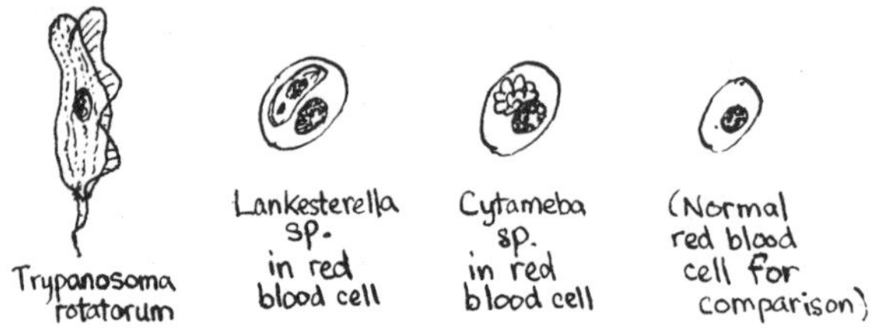

Intestine and rectum: Cut out a short portion of intestine and open it up with scissors; flush out its contents with a small amount of saline and set them aside. Now, gently scrape the lining of the intestine with a dissecting needle and rinse it with a few drops of saline; make slides of the rinsings. Examine the slides, using a low-power objective at first.

Make a slide of the intestinal contents also, teasing the mass apart with needles, if needed. Examine these slides also; refer to the protozoa shown on the next page.

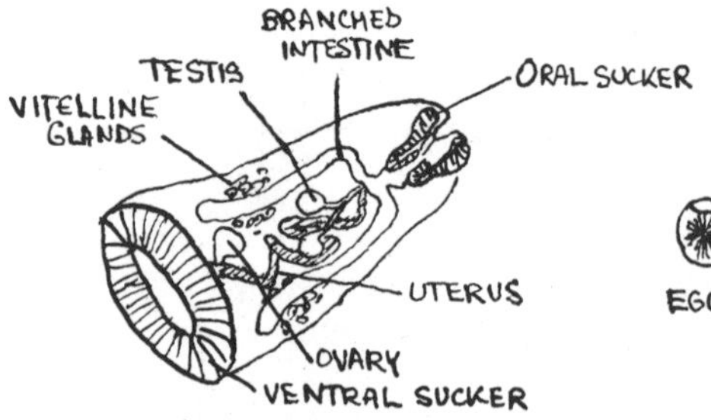

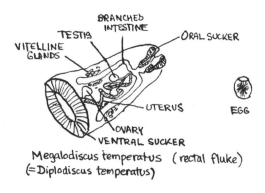

Megalodiscus temperatus (rectal fluke)
(=Diplodiscus temperatus)

Remove a bit of rectum and make slides from it as you did for the intestine. Examine the inner surface for flukes. The most common variety, *Megalodiscus temperatus*, is about 6 mm long. Its eggs pass out with the feces and are (of course!) eaten by snails. But in this case, after reproducing asexually in the snail, the larvae attach to the skin of tadpoles and frogs. When a frog sloughs its skin, it normally eats the sloughings, so the cycle is completed without a third host.

Make slides of the rectal contents as you did for the intestinal contents. You may see the following protozoans in these slides as well as those made from the intestinal contents.

The *Entameba ranarum* is similar to the organism causing amebic dysentery in man; it simply digests its way through the intestinal wall, phagocytosing red blood cells as it goes. Perforation of the intestine and infection of the abdominal cavity eventually result.

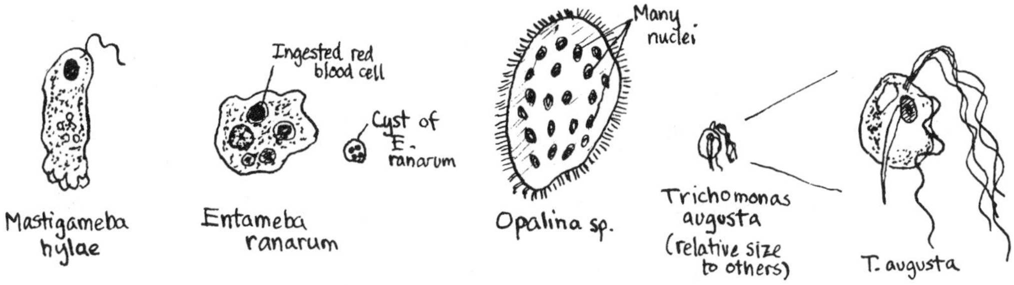

Questions:

1. Did your frog seem reasonably healthy? Were you surprised at finding so much life in it? What might you conclude about parasites in wild populations generally?
2. Some parasites feed and reproduce so much that they kill their hosts; some appear to do very little harm. Which would you judge to be better adapted? Often parasites of these extreme types are closely related. How might this be explained?
3. Did the flukes you saw seem to have any notable structural differences from other, nonparasitic flatworms? How can you account for them?

The world of symbiosis is much more complex than what we have seen here. For instance, there is evidence that the *Trychonympha* you saw *cannot* digest wood after all, but has symbiotic bacteria within it that do the job! One is reminded of the old rhyme—

Great fleas have little fleas
Upon their backs to bite 'em,
And little fleas have lesser fleas,
And so, *ad infinitum*.

Perception and Behavior

In most biology, we are concerned with the physical and chemical functions and composition of organisms, and many of the data are on one level obvious. Are you allergic to shrimp? Well, your body makes certain proteins that react with shrimp and set off a whole chain of nasty consequences. That is simple enough. But what about that *person* you can't stand (who may elicit symptoms as violent as the shrimp)? "Oh," says the biologist, with a wave of the hand, "that's *psychology!*" That is begging the question. Has it got a physical basis, or not? We really haven't the faintest idea, at least not for humans. But we can say that some behavior in animals definitely has a physical basis. Sometimes particular chemicals can be identified with particular kinds of behavior, or particular nerve circuits. Bit by bit, a picture is emerging.

One way to think about behavior is to compare it to a computer program. There must first be some kind of input, the computer works on it, and then there is an output. Some trigger in the environment—some cue perceived by the senses—sets off a pattern of behavior.

It makes sense, then, to consider perception and behavior together, and this is quite easily done with certain fundamental activities like mating and feeding. You can derive a lot of interesting ideas, for instance, from observing the aggressive displays of male Siamese fighting fish; these nasty-tempered little fish will react dramatically to their own reflection in a mirror, or to a rather abstract picture of themselves. You might spend several profitable hours trying to determine just what sets them off.

We will use less exotic creatures—blowflies. Their feeding behavior is quite predictable and has formed the subject of a classic exercise described by Vincent Dethier in his classic book, *To Know a Fly* (which should be required reading for anyone interested in biology). We will determine the limits of the fly's ability to detect sugar, the behavioral response made, how past experience alters this response, whether the fly can be fooled, and how our own sensitivity compares to the fly's.

First, you need to be able to control a fly. For this, you need a stick with a glob of sticky wax on one end. Dipping a wooden applicator stick in melted beeswax will do, or dental wax may be substituted.

Next, you must attach a healthy blowfly to the stick, and this takes some doing. You can put the fly in a jar with some ether-soaked cotton, or in a jar flooded with CO_2 gas; but take the fly out as soon as it ceases purposeful movement (i.e., its legs may still be kicking). If you leave it in any longer, it will never wake up.

A safer technique is to place the fly in the freezer for a minute or so—the chill will stun it, and this gives you enough time to work.

Now, *quickly* press both the fly's wings into the wax with a needle heated slightly in a flame. The needle need not be very hot, and it should *not* touch the fly. Try to get as large an area of the wing stuck to the wax as you can.

There is not likely to be a shortage of flies, so do several this way. They should emerge from anesthesia fairly quickly, unless you overdid it. Spend a few minutes looking closely at your captives—notice especially the mouthparts, retracted at present. (Beware—the blowfly is rather strong, and if your sticks are light, you may want to attach them to something!)

Now, obtain a petri plate full of distilled water and (working in a good light) lower your fly gently until its feet touch the water. Describe what happens. Can you see any accompanying abdominal movements? Here is a simple bit of behavior—the feeding response, made obvious in the fly by the lowering of the proboscis.

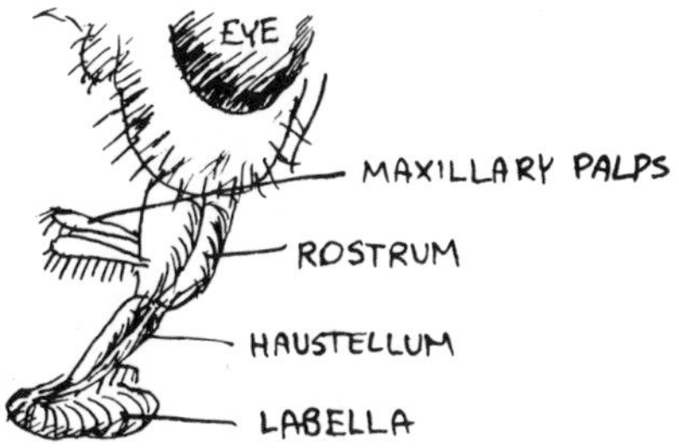

The labella are paired, padlike structures at the end of the proboscis; they are grooved and act like a sponge to sop up liquids. If your fly is thirsty, hold it over the water and let it drink. It will retract its proboscis when it has had enough.

Sugar, as any picnicker knows, is attractive to flies. And it does seem, sometimes, that they are uncannily sensitive to its availability. But just *how* sensitive are they? To find out, you need to make a number of different dilutions of sugar in water, and see how dilute a sugar solution the fly can detect.

Since you have no idea of the range of concentration involved, you want to do dilutions that cover a lot of ground quickly. These are "serial dilutions," a very simple method that is done as follows:

Take a 5% solution of sucrose (5 g in 100 mL water) and put 10 mL of it into a test tube. Label this 1. Put 5 mL of distilled water in each of nine additional test tubes. Then remove 5 mL of sucrose solution from the first tube and mix this with the 5 mL of water in the next tube (rinse the pipette with the mixture to get the last bit out). Label this tube $1/2$—you see why, of course. Now take 5 mL from the tube $1/2$ and mix this into the next tube, which you label $1/4$. Keep on like this, and you will soon have 10 tubes with labels running from 1 to $1/512$. In this last tube the concentration of sucrose is less than 10 mg per 100 mL. You can easily go farther (your $1/512$ tube will have 10 mL of solution in it).

Now, pour out some of the $1/512$ tube in one petri plate and some distilled water in another. Hold your fly over the water as before, until it stops drinking. Then hold it over the $1/512$ dilution of sucrose. If the proboscis is lowered, move the fly back to the plain water; the proboscis should retract. What does this demonstrate?

If the proboscis isn't lowered at $1/512$, pour that out and replace it with the next tube, $1/256$. Now what response do you get? Keep on working up the series until the fly feeds; then move it to water, as above.

If the fly feeds at $1/512$, and time permits, do another 10 tubes as before beginning with $1/512$. What concentration of sucrose do you have in the tenth tube now? Remember to keep giving the fly plenty of water.

Hold the fly over the first concentration of sucrose at which it feeds until it retracts its proboscis. Now move it to the next higher concentration; what happens?

The fly's sensitivity may be compared to a human's; taste the dilutions you have made, starting with the most dilute ($1/512$) and rinsing with plain water in between. At what dilution can you first say with confidence that you taste sugar? Perhaps you should let a partner give you the solutions with random samples of water in between, to counteract your imagination.

The results may come as a shock. Compared to flies, we are fairly insensitive. You might compile class data, because people will vary. Are there differences between smokers and nonsmokers?

Back to the flies. By this time, your first patient has probably gorged itself silly and will consequently be useless. Take a fresh fly and let it drink its fill of plain water. Now, try to determine what will make a fly *reject* sugar water. (Yes, there are limits, even for a fly.) Take a sugar solution (say, your $1/2$ dilution—you may need to make more) and add the following to different petri dishes of it: (*a*) a few drops of 5% acetic acid; (*b*) a few drops of 5% sodium chloride; (*c*) a few drops of 5% ammonium hydroxide.

Leave one petri dish untreated. Let the fly start feeding on this dish, then switch it to one of the above, smoothly and quickly. Does the proboscis retract?

You might set up a series of dilutions of salt, using the sucrose solution instead of water. How much salt does it take to get the fly to reject the solution? Can you taste that much salt?

Are there other substances that have this effect? Try alcohol, caffeine, and—of course—insect repellent.

Last, try to fool your fly. Move two petri plates together, the one with sugar water and the other with plain water. Lower the fly so that its feet touch the sugar water but its proboscis is over plain water. What happens? Substitute salt water for plain—does this make any difference?

Questions:

1. Why would a fly require a much more delicate chemical sensation than a human? How does each get its food?
2. Did the last experiment show that the fly tastes *only* with its feet, or what?
3. Answer the questions in the text.

Notes for Instructors

1. "Introduction to Dissection":

Fetal pigs—sometimes sold as "pig embryos"(!)—should be purchased in the largest size possible (about 30 cm) and should be double-injected. If you must economize, sacrifice injection before size. The new preservatives are better than formalin but can still be irritating; be sure students follow directions about rinsing specimens, and see to it that the room is very well ventilated. Supply cold cream and/or gloves.

Dissecting supplies: Dissecting pans filled with wax are best; the surface of the wax can be smoothed and cleaned by playing a Bunsen burner flame over it, as shown, or by autoclaving for 10 minutes at 15 pounds' pressure.

Most commercial dissecting kits are a waste of money. Purchase instruments separately:

a. fine scissors: *not* "student grade," which won't cut paper, but of "highest" or "surgical" quality—about 3 times the price but longer lasting.

b. fine-pointed forceps: avoid the ones made of sheet metal.

c. macramé pins (also called "T pins"): available cheaply by the box; distribute 6 or 8 to each student.

These are the essentials; it is better to buy a good quality of these and nothing else, rather than more but inferior instruments. (Single-edged razor blades are fine for scalpels.) Additional items, in order of importance, are:

d. dissecting needles: a pair of straight ones for each student; wooden handles are fine.

e. scalpel: replaceable blade, preferably a #4 handle and # 20 or # 21 blades. The plastic handles don't survive many blade replacements; buy metal ones.

f. heavy scissors: same comments as for fine ones, above.

If there is need for a case, these can be purchased empty; but cases seem to encourage students to put away wet and dirty instruments, and since they retard evaporation they encourage rust.

Solutions: To keep specimens fresh, moist, and supple, spray or sprinkle them with the following solution during long dissections and when storing specimens:

200 mL glycerol (glycerin)
300 mL tap water
0.5 g thymol (5-methyl-2-isopropyl-1-phenol)

2. & 3. "Organs and Systems in the Pig":

All supplies as above. Also, each 2 to 4 students should have a strong (10 ×) hand lens or (better) a dissecting microscope.

4. "The Microscope":

If at all possible, avoid sharing microscopes. This can be very discouraging and frustrating. Provide a good slide of stained, sectioned tissue like intestine for the setting-up process, and encourage its use each time the microscope is set up.

Supply lens paper and indicate the proper way to use it—i.e., not with direct finger pressure but as a rolled-up and torn-off wad, and with some lubricant, even just breath condensation.

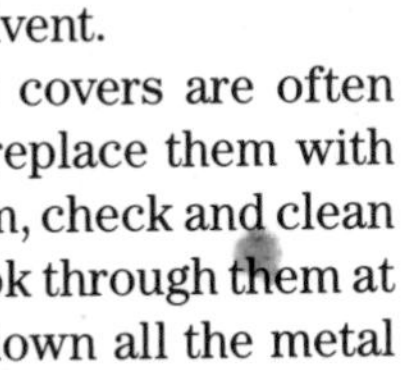

Note: Everybody knows about the danger of using xylene to clean a lens, but even alcohol will damage some coated lenses, so it is best to check with the manufacturer before using any solvent.

Note on microscope care: dust covers are often brittle and torn on older scopes; replace them with plastic bags. At the end of each term, check and clean the lenses (and be sure to really look through them at some standard object) and wipe down all the metal surfaces of the body and stage with a rag sprayed with one of the ordinary penetrating oil preparations (made for unfreezing rusted bolts). This will reduce the need for repairs and replacements.

Supplies:

a. for each student, a couple of slides and coverslips. Always see that these are clean, or have the student clean them.
b. prepared slides of intestine (cross-section), artery and vein, lung (alveoli), and connective tissue. These should all be mammalian, stained with hematoxylin and eosin. Prefer one set per student or pair of students.
c. A clean toothpick (flat kind) for each student.

Solutions: 0.1% methylene blue—note dye content on label, often around 80%; accordingly, make up 0.1%/0.8 solution, or 1.25 g in 1,000 mL distilled water. Filter before use. This is essential—the other 20% of the powder is insoluble gunk!

5. "The Protists":

Commercial cultures of *Paramecium sp.*, *Chilomonas sp.* (optional), *Euglena sp.*, *Ameba sp.*, and, if possible, a mixed culture. Methyl cellulose in dropping bottles.

Prepared slides of malarial blood smear (*Plasmodium vivax* or others) and of assorted diatoms.

Clean slides and coverslips, pasteur (disposable) pipettes and bulbs, microscopes.

Note: most protozoan cultures are sensitive to temperature shock—maintain them in a place free of drafts.

Yeast for phagocytosis: A half-hour before it is needed, make dyed yeast by mixing a packet of bakers' yeast with the following:

50 mL distilled water
5 g sucrose
0.5 g congo red (may substitute other dyes, such as methylene blue; but if congo red is used, students can observe pH changes during digestion—red in alkaline environments, blue in acid).

Just before use, dilute this with distilled water and allow the yeast to settle out (or use a centrifuge to spin them down); then pour off the liquid and replace it with 50 ml distilled water.

The standard protist-watching labs can be disastrous or delightful, and a lot depends on how familiar the instructor is with the care and feeding of the beasts. Here are some major points to watch in order to ensure success.

1. Fresh cultures are a must. Many cultures expire within a day or so of shipping. Order from the supplier nearest to you, and arrange for delivery the day of the lab. (Naturally, have a backup lab in case they don't arrive.) Immediately open and aerate the cultures as directed, but don't be so vigorous in pipetting them that you create a lot of bubbles. Many protists get trapped in the film of liquid on the surface of a bubble and expire there. As mentioned above, avoid temperature shocks. Also, avoid direct light (except for algal cultures) and excessive evaporation. Keeping the cultures in a plastic shoebox with wet paper towels in the bottom is often a good idea.

2. Handle the cultures carefully on the day of the lab. Don't stir them up or jolt them when you bring them into the lab (and caution the students not to, either); the jostling the critters get when put on a slide is quite enough, and much more may cause the more flexible ones to round up into an unrecognizable ball, and may break the flagella off the flagellated ones.

3. Check the cultures yourself just before lab. Some may be dead, and some may have been colonized by a beast other than what's on the label—usually you can put this to good use if you know about it beforehand.

4. Most important, advise students about the proper protist-hunting technique. The most common mistakes are:

a. Spastic pipetting: blowing air or water all over the bottom of the container dislodges the protists and makes it hard to find any. Squeeze the bulb *before* inserting it into the water.
b. Random grabbing: taking a sample by just sticking the pipette into the water anywhere doesn't work. How many students have you heard complain that they've never seen amebas, that these must be rare or delicate (or mythical) beasts? Yet they're really quite hardy. The problem is simply that they do not swim. They crawl *over surfaces* and therefore can only be caught by using the pipette like a vacuum cleaner on the bottom. Even the true swimmers can usually be captured best on the bottom, because that is where the food is. Teach students to go after the little bits of fluff on the bottom of the jar, and you'll hear far fewer cries of "I can't find anything!" in your labs.
c. Sloppy slidemaking: taking a 1-mL sample to make a slide is overkill. What usually ends up happening is that the student duly sucks up some fluff from the bottom, then follows it with a good pipetteful of clean water from above, and dispenses a drop of the latter on the slide, leaving the good stuff in the pipette. Take only a tiny bit—if you're using the disposable transfer pipettes of the usual sort

(pasteur pipettes), the narrow part of the pipette should not be half full. When the slide is made, you should be able to see some of the fluff on it, against a dark background. If it looks like clean water, it probably is.

Do make an effort to get pond water or aquarium filter gunk for the class. It is always fun to have the surprises that come with such samples, and it helps to dispel the impression that Paramecium is a synthetic commodity produced by the Giant Sloth Biological Supply Company specifically to torment students with. By the way, you will usually find rotifers in the filter water, and frequently some bizarre polychaete worms. This can lead to an interesting discussion of how you can tell something *is* a protist, especially considering that many of the rotifers will be dwarfed by ciliates and sarcodinids.

6. On Being a Metazoan:
Live *Hydra sp.* and *Daphnia sp.* to feed them; preserved *Grantia, Aurelia;* prepared slides of *Grantia*, c.s., and *Obelia*, w.m. Slides and microscopes. 5% acetic acid (5 mL glacial acetic acid, 95 mL tap water).

7. "The Worms That Got Organized":
Live *Planaria sp.;* live rotifers; live vinegar eels (*Turbatrix aceti*). Preserved tapeworms; depression slides, ice cubes, razor blades, small covered dishes such as 25 mm petri dishes. Hard-boiled egg yolk, dye (carmine or congo red).

8. "The Modular Approach":
Live earthworms (*Lumbricus cp.*), as large as possible; dissecting instruments, chloroform jars, preserved *Nereis virens.*

9. "The Armored Ones":
Dried or preserved horseshoe crabs (*Limulus sp.*) (these are for nondestructive study); large preserved crayfish (*Cambarus* or other)—don't bother with injected specimens. Dissecting instruments.

10. "The First to Fly":
Preserved *large* grasshoppers (*Romalea* or other); dissecting instruments.

11. "Happy as a Clam":
Preserved freshwater clam (*Unio* or other); preserved squid (*Loligo* or other); dissecting instruments.

12. "Strange Cousins":
Preserved starfish (*Asterias* or other); dissecting instruments.

13. "The Producers":
Cultures of *Oscillatoria, Chlamydamonas, Volvox, Rhizopus* (or grow your own bread mold!); bakers' yeast; prepared slides of dinoflagellates. Fermentation tubes (ungraduated), phenol red saturated in distilled water, sucrose, lactose, maltose, other mono- and disaccharides, also saccharin, other "sweeteners", all as 5% solutions. O.1 *N* HCl, O-1 *N* NaOH, methylene blue (from (4), above). Slides, coverslips, microscopes.

14. "The Terranauts":
Polytrichium sp., fresh preferably; fresh celery and/or daisies or other white flowers. Prepared slides of lichen thallus (c.s.), monocot and dicot stem (c.s.)—possibly also of leaf (c.s.). Fresh geranium leaf (or other fleshy leaf). Razor blades microscopes, slides and coverslips. A dye such as safranin, fuchsin, malachite green, or methylene blue.

15. "The Ubiquitous Ones":
Microscopes, slides, coverslips, methylene blue (from (4), above), toothpicks, Bunsen burners, and the following sterile supplies:

(*a*) nutrient agar: available powdered; make up as directed, preferably for a total of 25–50 mL per student (2–4 plates). Will require heating and stirring to dissolve. Then pour (while hot) into clean 100-mm petri plates (glass), half-filling each, and autoclave. (Can purchase ready-poured plates, or presterilized plastic plates, which can be filled with medium autoclaved in bulk.)

(*b*) swabs: purchase regular, sterile throat-culture swabs, or buy cotton-tipped applicator sticks (wooden handles only), wrap in foil, and autoclave.

(*c*) water: 300 mL tap water in a 1-L beaker, covered with foil and autoclaved.

All the above can be run together on a "slow exhaust" setting of the autoclave for 20 minutes at 20 lb. If autoclaving isn't possible, buy the supplies presterilized and boil the water for 1/2 hour.

For the Gram stain, prepare:

crystal violet solution A

crystal violet (powder) (= gentian violet)	2.0 g
ethyl alcohol, 95%	20.0 mL

crystal violet solution B

ammonium oxalate	0.8 g
distilled water	80.0 mL

Dissolve each of the above separately, combine all of both solutions, and filter before use. Keeps indefinitely.

Gram's iodine

iodine	1.0 g
potassium iodide	2.0 g
distilled water	300.0 mL

Grind the solids in a mortar and rinse the mortar with some of the water. Store out of light. Keeps indefinitely.

acetone-alcohol

ethyl alcohol, 95%	50.0 mL
acetone	50.0 mL

This keeps for some weeks, but should not be used when it is more than a few months old.

safranin

safranin (powder)	0.25 g
ethyl alcohol, 95%	10.0 mL

Dissolve stain in the alcohol, then add

distilled water	90.0 mL

Filter before using. Keeps indefinitely.

Identification of bacteria: There are, of course, thousands of possible types of bacteria your students could grow in their cultures. However, certain types are extremely common and can be counted on to show up in at least some cultures. They include

Staphylococcus: forms 2–3-mm wide, moist-looking colonies that are opaque off-white or sometimes yellow. Gram-positive cocci may be seen in bunches (this is best seen in broth cultures rather than plate cultures, however). The most common genus isolated from skin. It is most often (fittingly) *Staphylococcus epidermis*, although the occasional yellow colonies may be *Staphylococcus aureus*, which can be a pathogen.

Streptococcus: forms tiny colonies less than 1 mm wide, transparent, and barely off-white in color, and stains as a gram-positive coccus. The chains may be visible in direct mouth scrapings, but tend to disappear in plate cultures. Very common in nose and mouth. Species may include *Streptococcus salivarius* and *Streptococcus lactis*. A hemolytic species, *Streptococcus pyogenes*, is the cause of "strep throat."

Bacillus: forms large colonies over 3 mm wide, usually off-white, opaque, and often waxy-looking. Stains as large, gram-positive bacillus; spores may be visible as poorly stained oval swellings (and may be more common than cells in old cultures). Many possible species. One of the most common genera of decomposers.

Coliforms: typically 2–3-mm wide, transparent, off-white, and frequently evil-smelling colonies, quite wet-looking. Stains as small (almost too small to see), gram-negative rods. Common in human (and other mammals') intestines, and consequently in sewage and sewage-contaminated things. Often *Escherichia coli*, but includes a host of other genera and species. Their ability to ferment lactose (milk sugar) is characteristic of the group.

Pseudomonas: similar to coliforms but often forms characteristic yellow-green, green, or blue-green pigment, which diffuses into the medium. No other type does this. Although fairly common, these bacteria can produce severe infections, especially in burn patients.

Yeasts: form very large colonies over 4 mm wide, strikingly white and glistening, and very raised. Staining will show typical large, oval cells.

Molds: characteristic hairy-looking colonies, stains show hyphae (broken ends distinguish from bacilli, as does size).

The identification of their trapped "wild" bacteria can become a stimulating puzzle for students, and a surprising amount of classification can be done with minimal materials. A good microbiology laboratory manual should be consulted if there is interest in going farther.

16. "The Tools of the Trade":

For the size and distance estimate, you might go outdoors, or use a large, bare room. Any sort of stuffed specimens may be used, the more unusual-looking or brightly colored, the better. Control time very closely.

Microscope with disc-type ocular micrometer; stage micrometer; Wright-stained human blood smear. 1 set per 8–10 students.

Triple-beam balances (or whatever balances are available), 1 per pair. Small beakers (5–50 mL), pipettes (1–10 mL), water.

For spectrophotometry, any solution that absorbs in the visible region may be used; two standard ones are chromium (III) nitrate, 0.05 *M*, and cobalt (II) nitrate, 0.02 *M*. Both may be read at 500 mm.

17. "The Cell's Alchemists":

Potato prep: cut up a medium-sized white potato (peeled) in ½ L of water; puree in a blender and strain through cheesecloth. Keep the solution on ice.

Other solutions:

catechol, 1 g per 100 mL distilled water; keep on ice.
EDTA (use disodium salt) 0.5 g/100 mL distilled water, aid solution by dripping in 0.1 *N* NaOH.
PTU (0.5 g/100 mL distilled water)
acetic acid and NaOH, both 0.1 *N*.
pH paper, 37° water bath, boiling water, refrigerator and freezer.

Allow 20 mL per student of potato prep and catechol solution, 1 mL of others.

18. "The Master Molecule":

Calf thymus ("sweetbread") is the best tissue for this exercise; you can find it in many gourmet food stores. The best way to get it is directly from the slaughterhouse on the day the animal is slaughtered; if you cannot get there, ask that the thymus be frozen, not refrigerated, until you can pick it up. The tissue will undergo autolysis in a few days in the refrigerator, and although you will still be able to purify DNA from it, the DNA will be highly fragmented and will probably not spool out properly. If you see a fluffy precipitate that won't spool out, this indicates fragmented DNA.

On the other hand, the tissue keeps frozen for at least four years, in my experience, and can be repeatedly thawed and refrozen. Just don't keep it thawed for more than an hour or so before use.

Other tissues can, of course, be substituted for thymus, but the results are less satisfactory. Liver is the next best, and if at all possible obtain a fresh, whole one. Since small quantities are required, you may be able to use a liver from a freshly killed rat. Any tissue like liver must be perfused; inject physiological saline (0.85 g NaCl/L distilled water) with a syringe and needle into the larger blood vessels until the liver bleaches from dark red-brown to buff color. Also, use an amount of tissue that is about 4 times that called for below.

For every 50 students, take 10 grams of thymus and homogenize it in 200 mL of prep buffer; I have successfully used simply a mortar and pestle with a little clean sand added. If you have any glass tissue homogenizers available, or the Potter kind with teflon pestle and motor drive, these do give far better results. Kitchen blenders are quite acceptable if you take care not to overdo the blending. (Note that the thymus and prep buffer are nontoxic.)

As a first step, cut up the thymus with a scalpel (# 22 blade) or sharp knife. Try to produce chunks about 1 cm thick or less. Discard any obvious bits of connective tissue. Thymus tissue has a peculiar firm, smooth texture and an off-white color. As you cut the chunks, put them in chilled prep buffer on ice. During the cutting procedure, it is a good idea to wear rubber gloves and to use thoroughly clean instruments and glassware, preferably rinsed with alcohol or acetone and dried.

Homogenize the tissue on ice if possible; some glass homogenizers have hollow pestles you can fill with ice or ice water. If you are using a kitchen blender, stop frequently and put the pitcher part on ice for a few minutes. After several minutes of blending by any method, you will still have a considerable amount of connective tissue and debris, but there should be a completely opaque suspension of nuclei on top. Either strain the homogenate through about 4 layers of cheesecloth or spin it at about 300 × g (low speed in a tabletop centrifuge) to sediment the debris, and then use the supernatant in the next spin.

Spin down the nuclei at about 900 × g (nearly top speed in most table top centrifuges). If you have a refrigerated centrifuge, use it for this. A 4-minute spin should give you a substantial pellet of a fairly whitish color. This is the nuclear pellet.

Pour off the superantant, holding the pellet on the upper side of the tube, and resuspend it in fresh prep buffer. Replace the full amount of prep buffer, more or less; there is a great deal of leeway in the amount used.

Even if you elect not to have the students do the microscopic examination of the material, you should take a look at it after resuspending. There should be a number of distinctly stained, more or less round nuclei in each 40-power field of view. (The thymus cell nuclei are very small and are often distorted at this point into irregular shapes, but a smear of random-sized bits indicates lysed nuclei, which will probably not produce DNA large enough to spool out.)

For the spooling process, a little hook or knob is desirable on the tips of glass rods or pipettes used; a wood applicator stick also works, although if the absorbance readings are going to be done it may contribute some contaminants. It is simple to hold the tip of a disposable glass pipette in the flame of a Bunsen burner to slightly deform it; it is probably best to do this ahead of time, since alcohol will be used in the lab.

Prep Buffer:

57 g sucrose

3.1 g $MgCl_2 \cdot 6H_2O$

0.6 g Tris-HCl
(= Trishydroxymethylamine, or THAM; many firms have proprietary versions. Any grade is acceptable.)

Make up to 500 mL with distilled water.

Adjust the pH to 7.5 with 0.1 *N* HCl (the solution will normally start out with a high pH).

Store refrigerated. May be frozen for long-term storage. Keeps for at least a week refrigerated; discard if cloudy or moldy.

EDTA:

0.72 g disodium EDTA • $2H_2O$
(= Disodium ethylenediaminetetraacetate, versene)

Make up to 250 mL with distilled water.

Adjust pH to 7.5 with 0.1 *N* NaOH; this will help dissolve the EDTA. Check pH when completely dissolved.

Store at room temperature. Keeps indefinitely.

SDS:

25 g sodium dodecyl sulfate
(= sodium lauryl sulfate, SLS)

Make up to 250 mL with distilled water; dissolve gently, avoiding excessive foaming

Store refrigerated. Keeps indefinitely (will freeze solid in the refrigerator; simply thaw before use)

2 M *NaCl:*

29.2 g sodium chloride

Make up 59 250 mL with distilled water.

Store at room temperature. Keeps indefinitely.

Alcohol:

Normally, pure ethyl alcohol (100%) is used in this procedure. However, we have had perfectly good results with isopropyl alcohol, which may be easier to obtain in some situations. Even the 70% isopropanol sold as "rubbing alcohol" seems satisfactory. These alcohols can also be used to clean the glassware before the lab, although acetone will evaporate faster. Wait to use the glassware until it no longer smells of the alcohol or acetone.

Aceto-Orcein:

1 g orcein stain

55 mL acetic acid, glacial (i.e., concentrated)

Boil 5 minutes (in a fume hood) to dissolve the orcein powder. Cool to room temperature and add 45 mL lactic acid (or 45 mL distilled water; the lactic acid formula keeps longer).

Filter before use. Store at room temperature, not in direct sunlight. Keeps for years (if made with lactic acid). To use old stain, filter again before use.

This particular stain in not the only one that can be used. If methylene blue (see "The Ubiquitous Ones") is available, it can be substituted. In fact, almost any basic or neutral stain will do. Some stains may need to be diluted so that the background is not too dark.

Substitutions:

This procedure can be adapted for extremely tight budgets or difficulty with obtaining chemicals. It works with:

prep buffer: table sugar instead of sucrose
Epsom salts instead of $MgCl_2$
buffered aspirin instead of Tris-HCl

EDTA: may be omitted.

SDS: liquid Joy detergent instead of SDS

2 *M* NaCl: kosher salt (not iodized table salt)

alcohol: rubbing alcohol

aceto-orcein: methylene blue or crystal violet (= gentian violet) stains, made up in concentrated form, are used for topical treatment of fungal infections and can be bought in drugstores. Dilute them for use, typically about 1 to 10.

In each case, substitute the same weighed amounts as called for in the directions.

19. "Chloroplasts":

Live Anacharis (Elodea)—easily obtained in aquarium stores. Each student needs only a few leaves. Live spinach—the fresher, the better.

However, the plastic-wrapped spinach from the grocery store will work provided it is not becoming yellow or slimy. You will do well, however, to double the amount of spinach used in this case.

Extraction solution:

150 mL methanol, anhydrous, acid-free

50 mL petroleum ether

10% NaCl

50 g sodium chloride

500 mL distilled water

The extraction solution does not keep and should be made up fresh. Be especially careful to use the best grade of methanol you can obtain, since acid in this will destroy the pigments.

Do the extraction in a dim light (especially avoid fluorescent lights), using a 1-L separatory funnel and remembering to vent the funnel as soon as it is inverted.

Chromatography solvent

150 mL petroleum ether

10 mL acetone (acid-free)

The chromatography solvent does not keep and should be made up fresh. It is important to obtain the best grade of acetone you can, since acids in it will destroy the pigments. Use any chromatography paper you have, or substitute Whatman No. 1 filter paper cut to size.

The chromatography will go more rapidly and evenly if you put the solvent in the jars ahead of time and cover them to allow the atmosphere in the jar to become saturated with solvent vapors. Half an hour is usually sufficient time for this. Caution students not to open the jars until they are ready to put their chromatograms in.

An outer paper liner in the jar, soaked in the solvent, will protect the pigments from light and also help speed up the development.

A germicidal lamp or mineral light may be used as the U.V. source; short-wave U.V. is preferable.

Although there are relatively small amounts of solvents involved in this lab, it is best to work under a fume hood (if available) or in a very well-ventilated spot. Be sure to caution the students against looking directly at the U.V. lights.

Homogenize spinach leaves in 0.5 *M* sucrose, stain through cheesecloth, and spin down at $900 \times g$ for 10 minutes to pellet chloroplasts.

The Hill reagent may be 0.1% aqueous dichlorophenol indophenol (2,6 dichlorophenol indophenol) or 0.1% aqueous 2,3,6-trichlorophenol indophenol. Keep the reagents away from light and refrigerated; use only fresh reagents that have been properly stored, and make fresh reagent for each lab.

20. "The Cell Surface":

Live sponges; best are Red-beard (*Microciona prolifera*) and Yellow boring (*Cliona celata*), supplied by several companies in the southeast. Silk bolting cloth, 10 cm^2.

Artificial seawater—use commercial products.

Blood typing sera, lancets, swabs from commercial suppliers. Be sure that times indicated by serum supplier are adhered to, and that Rh sera are done on a warm surface.

For the sponge lab—gentle agitation with a rotary shaker or by hand will speed up the process.

21. "Mitochondria":

Succinate Dehydrogenase

Materials:

Mannitol grinding medium

(0.3 *M* D-mannitol in 0.02 *M* phosphate buffer, pH = 7.2)

54.66 g D-mannitol

1.98 g Na_2HPO_4

0.82 g KH_2PO_4

Make up to 1.0 L with distilled water.
Adjust pH to 7.2 with 1 *N* NaOH and refrigerate.

Assay Medium

(0.3 *M* D-mannitol in 0.02 *M* phosphate buffer, with 0.01 *M* KCl and 0.005 *M* $MgCl_2$)

To 500 mL of mannitol grinding medium, add
0.38 g KCl

0.51 g $MgCl_2 \cdot hH_2O$

Readjust pH to 7.2 if needed.
Dispense in 100-mL portions.

Azide

(0.04 *M* sodium azide)

0.26 g sodium azide

100.0 mL distilled water

Dispense in 10-mL portions.

DCIP

(5×10^{-4} M DCIP)

0.145 g 2, 6-dichlorophenolindophenol (Na salt)

1000.0 mL distilled water

Prepare immediately before use. Does not keep.
Dispense in 10-mL portions.

Malonate

(0.2 *M* sodium malonate)

3.32 g sodium malonate • $2H_2O$

100.0 mL distilled water

Adjust pH to 7.0 1 *N* HCl. Refrigerate. Dispense in 10-mL portions.

Succinate

(0.2 *M* sodium succinate)
5.40 g sodium succinate • $6H_2O$
100.0 mL distilled water

Adjust pH to 7.0 with 1 *N* HCl. Refrigerate. Dispense in 10-mL portions.

Supplies

2 fresh heads of cauliflower

2 scalpels with clean, sharp blades

2 mortars and pestles, chilled

25 g sand, chilled

cheesecloth

2 funnels

centrifuge tubes

5 ice buckets full of ice

1 boiling water bath with test tube rack

50 perfectly clean 13- × 100-mm test tubes

5 screw-cap test tubes

6 5- × 0.1-mL pipettes

25 1- × 0.1-mL pipettes

6 pipette bulbs or pipettors

Parafilm

Procedure:

Homogenization and centrifugation:

1. Cut off the outer 2–3 mm of the cauliflower florets and weight out 100 g of them.
2. Place these in a chilled mortar with 25 g of chilled, clean sand and grind them up for 4 minutes. (Split into 2 mortars if possible to speed up the work.) Add grinding medium as needed.
3. Filter the suspension through 4 layers of cheesecloth and squeeze out the juice. Place in centrifuge tubes and balance.
4. Spin at 600 × g for 10 minutes with refrigeration on.
5. Decant supernatant into clean centrifuge tubes and balance.
6. Spin at 10,000 × g for 30 minutes.
7. Decant supernatant and resuspend pellet in 35.0 mL of cold mannitol assay medium. (This is the total figure for mitochondrial suspension.)
8. Dispense 7.0-mL portions into sealed test tubes kept on ice—one tube for each team.

Oxygen Uptake

Use the assay medium and succinate solution from the section above. Make up cyanide and DNP solutions as below; these keep for at least a week if refrigerated.

Note: The cyanide solution is toxic. Although it is true that the entire amount called for in the lab is only a fraction of the lethal dose, even small amounts can damage tissues. Use pipetting aids and caution students to wash hands before using bathroom, eating, and so forth. DNP is toxic (less so than cyanide) and also stains clothing and skin. Avoid contact with this, too, and use care when weighing out the powder.

Sodium cyanide

sodium cyanide	0.049 g
assay medium	100 mL

(This is 10 m*M* cyanide.)

Refrigerate when not in use; keeps for a week. Avoid contact, ingestion, and acids. Highly toxic.

Dinitrophenol

dinitrophenol (DNP)	0.184 g
assay medium	100 mL

(This is 10 m*M* DNP.)

(The dinitrophenol called for is the 2,4-dinitrophenol isomer, the more common one. Avoid contact and ingestion. Highly toxic.) Refrigerate when not in use; keeps for a week.

Manometer fluid

Mineral oil or immersion oil can be used in the capillary tubes; add a few mg of Oil Red O or other oil-soluble dye to each 100 mL to give a color.

Don't use a water-based fluid in this experiment.

Supplies:

millimeter graph paper

clear plastic tape

scissors

5 microliter capillaries

13- × 100-mm test tubes, 5 for each student or group

stoppers to fit tubes, preferably soft rubber, 5

Parafilm (R) or similar laboratory film

test tube racks to securely hold 13- × 100-mm tubes

1.0- × 0.01-mL pipettes, 5 for each student or group

Pans of water or (better) a water bath set at some temperature around 30°. (Exactly what temperature doesn't matter, but it should be one the bath can maintain closely. With many baths you may need to go as high as 35°.)

22. "Muscle Contraction":

Several nice commercial preparations of glycerinated muscle, usually rabbit psoas muscle, are available. They keep (in the glycerol in the freezer) for years. Often the commercial kits come with some solutions prepared already; these do not keep very long (weeks at most) if refrigerated, and they provide too pat an answer for this approach. If at all possible, then, make up your own solutions for use with the muscle, as follows:

ATP	
adenosine triphosphate, trisodium salt	0.25 g
distilled water	100.0 mL
ADP	
adenosine diphosphate, disodium salt	0.21 g
distilled water	100.0 mL
GTP	
guanosine triphosphate, trisodium salt	0.26 g
distilled water	100.0 mL
$MgCL_2$	
magnesium chloride	0.095 g
distilled water	100.0 mL
$CaCl_2$	
Calcium chloride, anhydrous	0.111 g
distilled water	100.0 mL
KCl	
potassium chloride	0.75 g
distilled water	100.0 mL
NaCl	
sodium chloride	0.59 g
distilled water	100.0 mL

Other solutions as desired; an interesting thing might be to try the effects of 1 m*M* sodium lactate.

For the solution labeled "MAGIC," instructors are referred to the commercial kits or to the literature for the appropriate combination of the above solutions—since the author was fond of finding the answers in the back of the book when he was a student.

23. "Dissecting the Cell":

Sucrose for homogenization and centrifugation: 0.25 *M* sucrose, 5 m*M* $MgCl_2$

sucrose	8.5 g
magnesium chloride, $MgCl_2 \cdot 6H_2O$	1.0 g

TCA:	
trichloroacetic acid	10.0 g
distilled water	100.0 mL

orcinol:	
$FeCl_3 \cdot 6H_2O$	1.25 g
concentrated HCl	250 mL

Dissolve the ferric chloride in the hydrochloric acid, and then add

orcinol	2.5 g

This should be made just before use and not kept long. If it must be used for a week's worth of labs, refrigerate it.

diphenylamine:	
glacial acetic acid	300 mL
concentrated sulfuric acid	8.25 mL
diphenylamine	3.00 g

Use only fresh diphenylamine. This solution can be kept in the refrigerator for a week, but is best made fresh.

Biuret reagent:	
A. NaOH	60 g
distilled water to make 600 mL	
B. $CuSO_4 \cdot 5H_2O$	3.0 g
$NaKC_4H_4O_6 \cdot 4H_2O$ (sodium potassium tartrate)	12.0 g
distilled water to make 100 mL	

While stirring solution B, above, add solution A very slowly. When all is added, bring the volume of the solution up to 2.0 L with distilled water. Bottle in plastic and store at room temperature in the dark. The solution will keep, in clean bottles, for months at least. Discard if a dark precipitate is noted.

24. "The Dance of the Chromosomes":

Onion sets can be put out on hardware cloth supports, root ends in water, a week ahead of time.

Aceto-orcein as in (18), above, *or* commercial Schiff's reagent and 5 *N* HCl as in second procedure; can make a simple Schiff's reagent as follows:

Schiff's Reagent

0.025 g basic fuchsin

150 mL distilled water

3.0 g sodium hyposulfate ($Na_2S_2O_4$)

Filter; stores at room temperature for over a year.

Possibly also provide slides of whitefish blastula for animal mitosis.

Good onion root-tip squashes (or other squash preparations) can be made permanent by frosting the slide on dry ice, prying off coverslip, covering tissue with absolute ethanol, 2 changes, then xylene, 2 changes, and mounting a fresh coverslip with a permanent mounting medium.

25. "Shuffling Genes":

Live crickets (male), 1 per student. Rest as in (21), above.

26. "Genes in Human Populations":

PTC papers, also possibly PTU, other taste papers.

Poker chips, checkers, or other things of 2 colors for "selection" game.

27. "Getting It Together":

Live, fertile sea urchins. A dozen are enough for any class up to 100 or so—the problem comes in being sure of having fertile males *and* females. Ask for a supplier that actually checks fertility. Most will ship whatever species is fertile at the time. Not available in very cold weather, usually.

Maintain urchins for several days *damp*, but not submerged, in refrigerator. For longer times you need a salt-water aquarium.

Tuberculin syringe & needle and 0.5 *M* KCl, about 2 mL per urchin.

Artificial seawater as in (20), above.

28. "Symbiosis":

Live termites and live (preferably *not* laboratory-bred) grass frogs; small ones are better. Dissecting instruments, microscopes, slides.

29. "Perception and Behavior":

Blowflies can be commercially shipped as pupae, and will eclose in 1–2 weeks. Beeswax is best for attaching them, but double-stick tape is simpler. Need 5 g/100 mL solutions of sucrose, acetic acid, ammonium hydroxide, and sodium chloride, and at least 10 bottoms or tops of petri plates per student or pair.